SECOND EDITION

Statistics in Kinesiology

SECOND EDITION

Statistics in Kinesiology

William J. Vincent, EdD

California State University, Northridge

Human Kinetics

Library of Congress Cataloging-in-Publication Data

Vincent, William J
    Statistics in kinesiology / by William J. Vincent. -- 2nd ed.
      p.   cm.
    Includes bibliographical references and index.
    ISBN 0-7360-0148-4
    1. Kinesiology--Statistical methods.   2. Statistics.
   3. Kinesiology--Research--Methodology.   I. Title.
    QP303.V53   1999
    613.7--DC21   98-37457
                                              CIP

ISBN: 0-7360-0148-4

Copyright © 1995, 1999 by William J. Vincent

Acquisitions Editor: Loarn D. Robertson, PhD; **Developmental Editor:** Laura Casey Mast; **Managing Editors:** Lynn M. Hooper-Davenport and Jennifer Miller; **Assistant Editors:** Cynthia McEntire and Chris Enstrom; **Copyeditor:** Judy Peterson; **Proofreader:** Erin Cler; **Indexer:** Marie Rizzo; **Graphic Designer:** Keith Blomberg; **Graphic Artist:** Francine Hamerski; **Photo Editor:** Boyd LaFoon; **Cover Designer:** Jack Davis; **Photographer (interior):** Tom Roberts, except where otherwise noted. Photos on pages 17 and 149 by Chris Brown. Photo on page 53 by Barbara Rowland; **Illustrators:** Tom Janowski and Tom Roberts; **Printer:** United Graphics

Printed in the United States of America   10 9 8 7 6 5 4 3 2 1

Human Kinetics

Web site: http://www.humankinetics.com/

United States: Human Kinetics, P.O. Box 5076, Champaign, IL 61825-5076
1-800-747-4457
e-mail: humank@hkusa.com

Canada: Human Kinetics, 475 Devonshire Road Unit 100, Windsor, ON N8Y 2L5
1-800-465-7301 (in Canada only)
e-mail: humank@hkcanada.com

Europe: Human Kinetics, P.O. Box IW14, Leeds LS16 6TR, United Kingdom
(44) 1132 781708
e-mail: humank@hkeurope.com

Australia: Human Kinetics, 57A Price Avenue, Lower Mitcham, South Australia 5062
(088) 277 1555
e-mail: humank@hkaustralia.com

New Zealand: Human Kinetics, P.O. Box 105-231, Auckland 1
(09) 523 3462
e-mail: humank@hknewz.com

This book is dedicated to my eternal family:

Diana, Steven, Daniel, Susan, Gail, David, Nancy,
and all who have gone before or will come after.

Contents

Preface

I have taught a course at California State University Northridge in measurement and evaluation in kinesiology for 30 years. Over the span of my teaching career, this course has changed its title from Measurement and Evaluation in Physical Education to Measurement and Evaluation of Sport, Dance, and Exercise, to the current title, Measurement and Evaluation in Kinesiology. These evolving titles reflect the changes that have taken place in this field of study in recent years.

Kinesiology is defined as "the study of the art and science of human movement related to sport, dance, and exercise." It encompasses, therefore, much more than physical education, which is "the school program of instruction and participation in sport, dance, and exercise." Kinesiology includes physical education, exercise science (biomechanics, exercise physiology, and motor behavior), athletic training, adapted physical education, dance, sport psychology, sport history, sport sociology, sport management, and so on. Statistical techniques may be applied to any subject that produces data. This book will consider statistical techniques using data primarily from the kinesiological subdisciplines of physical education, exercise science, administration, and pedagogy. Because of the close professional relationship between kinesiology and leisure studies/recreation, some examples from the latter discipline are also included.

When I first taught this course, it was necessary to use statistics books from other fields like psychology or business. It was clear to me that a statistics book was needed with examples from the discipline of kinesiology so that students in this field could work with data and variables familiar to them. Hopefully, such a book will render the subject easier to comprehend and make the learning of statistical concepts more enjoyable and applicable. In addition, I have attempted to write in such a way that the mathematical concepts can be understood by a student with basic algebra skills.

This book represents a formal organization of the class notes I have developed over the years while teaching both undergraduate and graduate courses. During that time I discovered that certain teaching techniques and examples were effective in helping students understand the statistical concepts. I have included these techniques and examples in the book. I have also created a decision tree, found inside the back cover, which I hope will be helpful in making decisions about which analysis to use based on the type of data being analyzed.

Any science-of-statistics teacher knows that before a computer is valuable to a student, the student must be able to solve problems with simple data by hand. Computers can do two things better than humans: they can remember more detail, and they can work faster. But they cannot think. The computer should be an extension of the human brain, where ideas flow from the brain, down the arms, through the fingers, and onto the keyboard. In this way, the advantages of the human brain and the speed and memory capacities of the computer are combined into a powerful processing and analyzing tool. The mind conceptualizes and the computer executes.

However, without an understanding of the formulas used by the software and a feeling for how to interpret the results, students tend to accept any result that the computer produces without critical evaluation. This book is written, therefore, with examples that students can first learn to solve by hand. Computers are wonderful machines. They are most useful, and they should be integrated into any course in statistics. But without an understanding of the underlying statistical concepts, the student may use the computer to produce incorrect results or be unable to interpret the answers the computer produces.

This second edition contains many editorial changes to make the text more clear to the beginning student. In the second edition, comments from instructors and students who used the first edition have guided rewriting so that concepts that were not clear in the first edition are elucidated.

The second edition contains a new chapter (11) on factorial analysis of variance, new sections on factor analysis and discriminant analysis in chapter 12, and new sections on Kruskal-Wallis ANOVA for ranked data and Friedman's two-way ANOVA by ranks in chapter 13. Students have consistently requested more opportunities to practice their calculation skills so additional problems have been included at the ends of selected chapters. In addition, at the request of students, I have edited some sections to make them more clear. References to methods of computer applications have been included where appropriate.

I appreciate the efforts of my students and others who have found errors in the first edition and have pointed them out to me. In my classes, I offered $1.00 to any student who found a typographical error and $2.00 for a mathematical error. I passed out the money at the beginning of each class based on the reading for the previous week. At the end of the first semester I was broke, but the students learned to read critically. These errors were recorded and have been corrected in the second edition.

In the preface to the first edition, I acknowledged many people who helped with this project. I am still indebted to them and once again express my thanks. But in the second edition, extra thanks must go to my students who have inspired me and encouraged me to complete this revision. They are the ones who have challenged me and provided the motivation to teach and to write. I have recently become an administrator, and I can honestly say that the classroom is where I find my rewards. It is there that the interaction takes place that inspires teachers and empowers learners.

Acknowledgments

I would like to thank the publishers and authors who have given me permission to reprint tables. Special thanks must be given to the late Professor E.S. Pearson of the University College in London and to the Biometrika Trustees. Professor Pearson responded personally to my request to reprint selected tables. His handwritten letters to me are cherished and valuable documents.

Special thanks to some of my colleagues at California State University Northridge are also in order. Linda Fidell and Barbara Tabachnick of the Department of Psychology have helped me immeasurably. I have audited their classes and sought help from them on many occasions. Much of what I know about multiple regression and multiple analysis of variance I learned from them.

I would also like to thank Diane Swartz and John Motil in the Computer Science Department for getting me started on elementary programming. My experiences in their classes have been most rewarding.

List of Key Symbols

α	1. Greek letter, lowercase *alpha*
	2. Area for rejection of H_0 on a normal curve
ANCOVA	Analysis of covariance
ANOVA	Analysis of variance
β	Slope of a line; Greek letter, lowercase *beta*
B	Greek letter, uppercase *beta*
χ	Greek letter, lowercase *chi*
χ^2	*Chi*-square
C	1. Column
	2. Number of comparisons
	3. Constant
	4. Cumulative frequency
	5. *Y*-intercept of a line
d	Deviation (the difference between a raw score and a mean)
df	Degrees of freedom
D_1	The 10th percentile
E	Expected frequency
ES	Effect size
f	Frequency
F	Symbol for ANOVA
FW_α	Familywise *alpha*
H	The highest score in a data sheet. Also, the value of Kruskal-Wallis ANOVA for ranked data.
H_0	The null hypothesis
H_1	The research hypothesis
HSD	Tukey's honestly significant difference
i	Interval size in a grouped frequency distribution
I	Scheffé's confidence interval
IQR	Interquartile range
k	Number of groups in a data set
L	The lowest score in a data sheet
MANOVA	Multiple analysis of variance
M_G	The grand mean in ANOVA
MS	Mean square
MS_E	Mean square error

μ	1. Greek letter, lowercase *mu*
	2. Mean of a population
n	Number of scores in a subgroup of the data set
N	Total number of scores in a data set
O	Observed frequency
ω	Greek letter, lowercase *omega*
ω^2	*Omega* squared
p	1. Probability of error
	2. Proportion
P	Percentile
Q_1	The 25th percentile
r	Pearson's correlation coefficient
R	1. Range
	2. Rows
	3. Multiple correlation coefficient
R_1	Intraclass correlation
R_2	Coefficient alpha in intraclass correlation
ρ	1. Greek letter, lowercase *rho*
	2. Spearman's rank order correlation coefficient
SD	Standard deviation (based on a sample)
SE_D	Standard error of the difference
SE_E	Standard error of the estimate
SE_M	Standard error of the mean
σ	1. Greek letter, lowercase *sigma*
	2. Standard deviation (based on a population)
σ_p	Standard error of proportion
SS	Sum of squares
Stanine	Standard score with middle score $= 5$ and $R = 1$ to 9
Σ	1. Summation
	2. The sum of a set of data
t	Student's *t*
T	T score (standard score with $\overline{X} = 50$ and $\sigma = 1.0$)
U	Mann-Whitney *U* test
V	Variance
X	A single raw score
$\overline{X}$	The mean
X_{mid}	The middle score in an interval of scores
Z	Z score (standard score with $\overline{X} = 0$ and $\sigma = 1.0$)
Z_α	Alpha point

Chapter 1

Measurement, Statistics, and Research

The most important step in the process of every science is the measurement of quantities. . . . The whole system of civilized life may be fitly symbolized by a foot rule, a set of weights, and a clock.

James Maxwell

All science is based on precise and consistent measurement. Whether or not you consider yourself to be an exercise scientist or a physical education teacher or a coach, there is much that you can gain by learning proper measurement techniques. Exercise scientists measure various attributes of human performance in laboratories, and teachers and coaches measure students' performances by giving them tests.

Maxwell's observation indicates that most measurements are of quantitative values—distance, force, and time. However, counting the frequency of occurrence of events is another way to measure quantitative values.

These same measurements apply to kinesiology, where we commonly measure *distance* (how tall people are, or how far they can jump), *force* (how much they weigh, or how much they can lift), *time* (how fast they can run, or how long they can run at a given pace on a treadmill), and *frequency* (how many strides it takes to run 100 meters, or how many times the heart beats in a minute). These measurements are sometimes referred to as *objective* because they are made with mechanical instruments, which require minimal judgment on the part of the investigator.

Other measurements, classified as qualitative (or subjective, because they require more human judgment), are used to determine the quality of a performance, such as a gymnastics routine, a golf swing, or a swimming stroke.

What Is Measurement?

Put simply, **measurement** is the process of comparing a value to a standard. For example, we compare our own weight (the force of gravity on our body) to the standard of a pound or a kilogram every time we step on a scale. When a physical education teacher tests students in the long jump, the process of measurement is being applied. The standard to which the jumps are compared is distance (feet and inches, or meters). When the teacher uses an instrument (in this case, a tape measure) to determine that a student has jumped 5.2 meters, this bit of information is called data.

Data are the result of measurement. When individual bits of data are collected they are usually disorganized. After all of the desired data are known, they can be organized by a process called statistics. **Statistics** is a mathematical technique by which data are organized, treated, and presented for interpretation and evaluation. **Evaluation** is the philosophical process of determining the worth of the data.

Precision in measurement is essential. If measurement is not precise, the results cannot be trusted. The data from measurement must also be reproducible—that is, a second measurement under the same conditions should produce the same result as the first measurement. Data that are not precise and not consistent are of no value. They may not represent the true conditions of the subjects being measured.

To be acceptable, data must be valid, reliable, and objective. **Validity** refers to the soundness or the appropriateness of the test in measuring what it is designed to measure. Validity may be determined by a logical analysis of the measurement procedures or by comparison to another test known to be valid. **Reliability** is a measure of the consistency of the data, usually determined by the test-retest method, where the first measure is compared to a second or third measure on the same subjects under the same conditions. **Objectivity** means that the data are collected without bias by the investigator. Bias can be detected by comparing an investigator's scores against those of an expert or panel of experts.

The Process of Measurement

Measurement involves four steps:

1. The object to be measured is identified and defined.
2. The standard to which the measured object will be compared is identified and defined.
3. A comparison of the object to the standard is made.
4. A quantitative statement is made of the relationship of the object to the standard.

For example, if we measured the height of a person who is 2 meters tall, we would conclude that the person's height (the object measured) is two times greater (the relationship) than 1 meter (the standard).

The standard used for comparison is critical to the measurement process. If the standard is not consistent, then each time an object is compared to that standard, the data will change. In the English system of measurement, the original standard was not consistent. About 600 years ago in England the standard for measuring distance was the length of the king's foot. When the king died, and another with a smaller or larger foot took his place, the standard changed, and the length of all objects in the kingdom had to be redetermined.

The English system of measuring distance was originally based on anatomical standards such as foot, cubit (distance from the elbow to the finger tips—typically about 1.5 feet), yard (a typical stride for a person), and hand (the spread of the hand from end of little finger to end of thumb). Force was based on pounds (7,000 grains of wheat equaled 1 pound), and each pound was further divided into 16 ounces. In years when the rain was adequate, the grains of wheat were large, but in drought years the grains were smaller. So the standard changed from year to year.

Eventually these measures were standardized, but the English system is difficult to use because it has no common numerical denominator for all units of measurement. Sixteen ounces make a pound, but 2,000 pounds make a ton; 2 cups make a pint and 2 pints make a quart, but it takes 4 quarts to make a gallon. Twelve inches make a foot, 3 feet make a yard, and 5,280 feet constitute a mile. It is no wonder that children have difficulty learning this system.

The *metric system*, which is more consistent and easier to understand, was first introduced in France in 1799. It is now used everywhere except in the United States and a few countries from the British Empire. In this system, the units of measurement for distance, force, and volume are based on multiples of 10. The metric system uses the following terms:

Basic Terminology in the Metric System

Prefix	Value
milli	1/1,000
centi	1/100
deci	1/10
deca	10
hecto	100
kilo	1,000
mega	1,000,000
giga	1,000,000,000

These terms make it easy to convert units of measure. For example, a kilometer is 1,000 meters, and a centimeter is 1/100 of a meter. The same terminology is used for force, volume, and distance. The metric system is based on geophysical values that are constant. A meter is 1/10,000,000 of the distance from the equator to a pole on the earth. A gram is the force of gravity on 1 cc of water at 4° centigrade (its maximum density).

Fortunately, both the English and metric systems use the same units of measurement for time: seconds, minutes, hours, and so on. Although measurements are not based on multiples of 10 (there are 60 seconds to 1 minute), the relationships are the same in both systems and are common throughout the world.

The metric system is clearly superior to the English system. If it were used everywhere in the world, there would be less confusion and frustration when students learn measurement concepts.

Variables and Constants

When we measure human performance, we measure variables. A **variable** is a characteristic of a person, place, or thing that can assume more than one value. For example, one characteristic of a person that varies is the time he or she takes to run a mile.

Other examples of variables in kinesiology are weight, height, the number of basketball free throws made out of 10 tries, oxygen consumption and heart rate during a treadmill test, angles at the knee joint during various phases of a running stride, scores on skills tests, and placement in a ladder tournament.

Note that individual people score differently on the same variable, and a single person may perform differently when measured on the same variable more than once. The data vary between people as well as within people. Because variables change, we must monitor, or measure, them frequently if we need to know current status of the variable.

A characteristic that can assume only one value is called a **constant**. Since a constant never changes, once we measure it with accuracy, we do not have to measure it again. The number of players on an official baseball team is a constant. It must be nine, no more or no less. The distances in track events are constants. In a 100-meter dash, the distance is always 100 meters. The distances from base to base on a baseball field and from pitcher's mound to home plate are constants.

Many anatomical characteristics are constants. Generally a person has only one head, one heart, two lungs, and two kidneys. Of course, illness or accidents may disfigure the body and thus change some of these characteristics, but typically the anatomy of the human body is constant.

Probably the most well-known constant in mathematics is pi. Pi is the number of times the diameter of a circle can be laid around its circumference, and it is always 3.14159. . . .

Types of Variables

Variables may be classified as continuous or discrete. A **continuous variable** is one that theoretically can assume any value. Distance, force, and time are continuous variables. Depending on the accuracy of the measuring instrument, distance can be measured to as short as a millionth of a centimeter or as long as a light-year. Time can be measured in milliseconds, days, or centuries.

The values of **discrete variables** are limited to certain numbers, usually whole numbers or integers. When counting people, we always use whole numbers because it is impossible to have a fraction of a person. The number of heartbeats in a minute and the number of baskets scored in a basketball game are also examples of discrete variables. One cannot count half of a heartbeat or score part of a basket.

Classification of Data

Data collected on variables may be grouped into four categories, or scales: nominal, ordinal, interval, and ratio. **Nominal scales** group subjects into mutually exclusive categories. There is no qualitative differentiation between the categories. Subjects are simply classified into one of the categories and then counted. Data grouped this way are sometimes called **frequency data** because the scale indicates the frequency, or the number of times an event happens, for each category. For example, a teacher classified students as male or female and then counted each category. The results were 17 males and 19 females. The values 17 and 19 represent the frequencies of the two categories, male and female.

Some nominal scales have only two categories, such as male or female, or yes and no. Others, such as an ethnicity scale, have more than two divisions. Nominal scales do not place qualitative value differences on the categories of the variable.

An **ordinal scale,** sometimes called a rank order scale, gives quantitative order to the variables, but it does not indicate how much better one score is than another. In a physical education class, placement on a ladder tournament is an example of an ordinal scale. The person on top of the ladder tournament has performed better than the person ranked second, but there is no indication of how much better. The top two persons may be very close in skill, and both may be considerably more skilled than the person in third place, but the ordinal scale does not provide that information. It renders only the order of the players, not their absolute abilities. There may be unequal differences between the positions on an ordinal scale. If 10 people are placed in order from short to tall, then numbered 1 (shortest) to 10 (tallest), the values of 1 to 10 would represent ordinal data. Ten is taller than 9, and 9 is taller than 8, but the data do not reveal how much taller.

An **interval scale** has equal units, or intervals, of measurement—that is, the same distance exists between each division of the scale—but there is no absolute zero point. Because zero does not represent the absence of value, it is not appropriate to say that one point on the scale is twice, three times, or half as large as another point. To determine if a scale is interval in nature we ask, Is it possible for a score to be less than zero? If the answer is yes, then the scale is an interval scale.

The Fahrenheit scale for measuring temperature is an example of an interval scale: 60 degrees is 10 degrees hotter than 50 degrees and 10 degrees cooler than 70 degrees (the intervals between the data points are equal), but 100 degrees is not twice as hot as 50 degrees. This is because 0 degrees does not indicate the complete absence of heat. In athletics, interval scores are used to judge performances in sports like ice skating, gymnastics, diving, and synchronized swimming. A 9.0 in gymnastics is halfway between 10.0 and 8.0, but it is not necessarily twice as good as a 4.5. A score of 0 does not mean the absence of skill; it means that the performance did not contain sufficient skill to be awarded any points.

The most complete scale of measurement is the **ratio scale.** This scale is based on order, has equal distance between scale points, and uses zero to represent the absence of value. All units are equidistant from each other, and proportional, or ratio, comparisons are appropriate. All measurements of distance, force, or time are based on ratio scales. Twenty pounds is twice as heavy as 10 pounds, and 100 feet is twice as far as 50 feet. A negative score is not possible on a ratio scale. A person cannot run a race in negative seconds, weigh less than 0 pounds, or score negative points in basketball. In kinesiology, data of the ratio type are often used.

Nominal and ordinal scales are called nonparametric because they do not meet the assumption of normality. Interval and ratio scales are classified as parametric. These concepts are discussed further in chapter 13.

Research Design and Statistical Analysis

Research design and statistical analysis are both based on measurement principles. They are so intertwined that it is difficult to discuss one without referring to the other. This book presents statistical techniques that are designed to assist in evaluating data. How the data are collected, which instruments are used, and how the variables are controlled are all part of the research design.

Research is a special technique for solving problems. Identifying the problem is a critical part of research. A problem usually begins with such questions as "I wonder if . . . ," "I wonder why . . . ," or "I wonder what . . . ." If we do not find the answer to the question, then we have a problem to be researched. First, we might ask an expert in the field about the question. If the expert knows the answer, our problem is solved.

If the expert does not know, we might visit the library. If we find the answer in the library, our problem is solved. But when we cannot find the answer by looking in the library or talking to other people, then the only way to solve the problem is to conduct research.

Many people think of an experiment when they hear the word research, but research is not limited to exploring scientific, experimental design problems. The problem may be solved by historical research, descriptive research, or experimental research. **Historical research** is a search through records of the past to determine what happened and why. It is an attempt to solve present problems by learning from the past. This book does not address historical research. Many texts on research design are available that discuss historical research.

Descriptive research involves describing current events or conditions. The most common tool of descriptive research is the survey. The researcher identifies the events or conditions to be described and seeks information from people or other sources by asking questions, often by using a questionnaire. Statistical techniques are used to organize, treat, and present the data from descriptive research for evaluation and interpretation.

Experimental design is the research process that involves manipulating and controlling events or variables to solve a problem. The remainder of this chapter discusses research of the experimental design type.

To begin our research, we need a plan. The research design is the plan that sets out the manner in which the data will be collected and analyzed. Often the plan calls for subproblems to be solved first or for pilot work (preliminary data collection) to be conducted to determine the feasibility of the research design. It is a common error of beginning researchers to jump into data collection before doing adequate planning and pilot work.

After the problem has been identified and the library search completed, if the problem has not been solved, we are ready to make a hypothesis. A **hypothesis** is an educated guess or a logical assumption that is based on prior research or known facts and that can be tested by the experimental design. The hypothesis must be stated in such a way that statistical analysis can be performed on the data to determine the probability (or odds) of obtaining the given results if the hypothesis is true.

Hypothesis Testing

The hypothesis that prompts the research is called the **research hypothesis** and is symbolized by H_1. It usually predicts relationships or differences between or among groups of subjects. After all, if we didn't believe that relationships or differences existed, then we would probably not perform the experiment in the first place. However, H_1 is not the hypothesis that is normally tested by statistical analysis.

The hypothesis to be tested statistically is called the **null hypothesis,** or H_0. The null hypothesis predicts no relationship or no difference between the groups. If H_0 is true, then H_1 is false, and vice versa. They are mutually exclusive propositions. Essentially, the null hypothesis states that any relationship or difference that may be observed by measurement is the result of random occurrences.

The statistical analysis reports the probability that the results would occur as we observe them if the null hypothesis is true. For example, the statistics may indicate the probability that the null hypothesis is true is less than 1 in 100. If the probability is small that H_0 is true, we reject the null hypothesis and accept the alternate, or research hypothesis (H_1). If the probability is large that H_0 is true, we accept the null hypothesis.

The null hypothesis is normally used to evaluate results because researchers are usually not sure what the results will be when they conduct an experiment. If we knew the results before we started, we would not need to do the experiment. But researchers do begin an experiment with an idea in mind, and this idea is represented by the research hypothesis. Usually the research hypothesis prompts the experiment, but it is the null hypothesis that is tested.

The researcher should test the null hypothesis unless strong evidence—from other research, from related literature, from the opinion of experts, or from other reliable sources—suggests that a significant difference is to be expected. The opinion (perhaps biased) of the investigator is not sufficient reason to test the research hypothesis.

We reject H_0 and accept H_1 when differences or relationships among variables are established beyond a reasonable doubt, or at an acceptable level of confidence. Before we collect data, we must establish a level of confidence. For example, we may decide to reject the null hypothesis if the odds against it are better than 95 to 5. Stated another way, p (the probability that the null is true) is $\leq .05$ ($p \leq .05$). This means that the odds of H_0 being true are less than or equal to 5 in 100.

Independent and Dependent Variables

An **independent variable** is one that is totally free to vary by itself and does not covary with another variable. Changes that occur in the first variable are not necessarily related to changes in the second variable. For example, if a person improves his or her balance in gymnastics, that has no relationship with or effect on that person's ability to consume oxygen (VO_2). Likewise, an improved VO_2 through aerobic conditioning has no effect on a person's ability to perform a handstand. Each variable (balance and oxygen consumption) is independent and is not affected by changes in the other.

A **dependent variable** is not free to vary. It is dependent on the effects of one or more other variables. For example, weight is partially dependent on height. During the growing years, as people grow taller, they also grow heavier. In adults, we generally find that tall people are heavier than short people. But height is not dependent on weight. A person can gain or lose weight with no effect on height.

Skill is dependent on practice. Usually the more one practices, the better one performs. If a graph of practice time (on the X-axis) and skill (on the Y-axis) were constructed, the plot of the relationship would progress from lower left to upper right (see figure 1.1). As practice time increases on the X-axis, skill would also increase on the Y-axis.

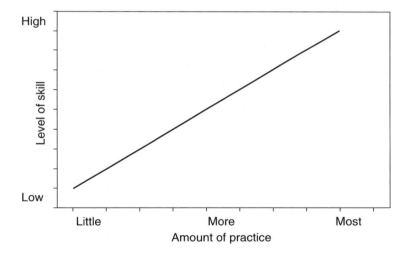

Figure 1.1   Relationship of skill to practice.

In experimental design, it is common to refer to the **independent variable,** which is controlled by the researcher, as the **predictor variable** and to call the **dependent variable,** which is being studied, the **criterion variable.** If we measured the effect of a diet on weight gain or loss, the diet would be the independent, or predictor, variable, and weight would be the dependent, or criterion, variable. The weight gain or loss is dependent on the effects of the diet. We manipulate the predictor variable and measure its effect on the criterion variable.

Validity

Research conducted by experimental design must demonstrate both internal and external validity. **Internal validity** refers to the design of the study itself; it is a measure of the control within the experiment to ascertain that the results are due to the treatment that was applied. Sometimes when people take motor skill tests, they improve their performance simply by taking the test twice. If a pretest-treatment-posttest design is conducted, the control subjects may show improvements on the posttest because they learned specific test-taking techniques while taking the pretest. If these improvements are attributed to treatment, when in fact they are due to practice on the test, an error has been made.

To find out if the changes are due to the treatment or to practice on the test, the researchers could use a control group. The control group would also take both the pre- and posttest but would not receive the treatment; the control subjects may or may not show posttest improvement. Analyzing the posttest differences between the experimental group and the control group helps us sort out how much improvement is due to (a) the treatment and (b) the learning effect from the pretest. A design of this type without a control group can be criticized for having weak internal validity.

Other factors, which are not controlled in the experiment, can also reduce the internal validity. These factors are sometimes called **intervening variables,** or extraneous variables; they intervene to affect the dependent variable but are extraneous to the research design. For example, if the posttest is given too soon after treatment, fatigue may affect the results. If fatigue is not one of the variables being studied, and if it is not properly controlled, the results may not reflect the real effect of the treatment. Likewise, a study of the effects of a low-fat diet on body composition that does not control for the variable of exercise may reach erroneous conclusions. To preserve internal validity, all potential intervening variables must be controlled or equated between the groups in the experiment.

Two other factors that may reduce internal validity are instrument error and investigator error. **Instrument error** refers to incorrect data that is due to a faulty instrument. If we use an inaccurate stopwatch to measure time, then the data will be faulty. Complicated instruments, such as oxygen analyzers, respirometers, force plates, and dynamometers, need to be continually calibrated to ensure that they are correctly measuring the variable.

Investigator error occurs when the investigator introduces bias in recording the data. This bias may be intentional or unintentional. Of course, we would hope that no one would intentionally falsify data, but it has been known to happen. Usually, however, the bias is unintentional.

For measurements of distance, force, time, or frequency, investigator error is usually minimal. It is fairly easy to read an instrument, especially one with a digital readout. But when judging a skilled performance, the investigator, or judge, can produce biased scores. Other errors occur when the investigator is not skilled in the data collection technique, such as taking skinfold measurements for body composition analysis. The amount of error an investigator introduces can be determined by comparing the investigator's data with that collected on the same subjects by an expert or panel of experts.

External validity refers to the ability to generalize the results of the experiment to the population from which the samples were drawn. If a sample is not randomly drawn, then it may not represent the population from which it was drawn. It is also possible that other factors (intervening variables) that were controlled in the experiment may not be controlled in the population. The very fact that the experiment was tightly controlled may make it difficult to generalize the results to an actual situation where these variables are left free to influence performance.

When conducting research we need to be able to demonstrate that we have designed the experiment properly and controlled all appropriate variables to assure internal validity. We also need to show that the results can be applied to the real world. The process of generalizing from a sample to a population is called **statistical inference** and is one of the basic tools of the statistician.

Statistical Inference

Much of the work performed by a statistician involves making predictions about a large group (usually a group of people, but it could be a group of animals, rocks, fish, stars, or almost any object) based on data collected from a small portion of the group. A **population** is any group of persons, places, or things that have at least one common characteristic. These characteristics are specified by the definition of the population. All of the seventh-grade girls in a given school would be one population. Males between the ages of 31 and 40 who do not participate in regular exercise are another population. In short, a population can be any group, as long as the criteria for inclusion in the group are defined so that it is clear who qualifies as a member.

Usually, the population of interest is quite large, so large that it would be either practically impossible or financially unreasonable to measure all of the members. If it is impossible or impractical to measure all members of a population, then we measure a portion, or fraction, of the population, which is called a **sample.** We assume that the subjects in the sample represent, or have the same characteristics as, the population. Thus data collected on the sample can be generalized to estimate the characteristics of the population.

The larger the sample, the more accurately it will represent the population. If we want to know the average height of all 15-year-old females in a city, we are more likely to get an accurate picture of the population by measuring 1,000 subjects than by measuring only 10 subjects. The error in the prediction is inversely related to the size of the sample. The larger the sample, the smaller the error, and the smaller the sample, the larger the error.

A **random sample** is a sample wherein each member of the population has an equal opportunity of being selected into the sample. Any inference made about the population based on a nonrandom sample is of doubtful value. In fact, all of the formulas and calculations that are used to make inferences about populations are based on the assumption of random sampling. Major errors in the estimate of the population can be made if the sample is not random.

For example, suppose we needed to estimate the average percent body fat of all males at a university. A random sample would require that all male students have an equal chance of being selected. If we select only males who enter the gymnasium between 1:00 P.M. and 3:00 P.M., the sample is likely to represent a larger portion of athletes than is in the population, because many athletes enter the gymnasium at this time to prepare for practice. Athletes tend to have low body fat, so such a sample would probably underestimate the average percent body fat of the entire population and would be considered biased.

Possibly there is no one place on campus at any given time where everyone has an equal chance of being chosen. If the sample is taken during the day, night students are excluded. If it is taken on Tuesday, then students who are not on campus on Tuesday are excluded. If it is taken near the engineering building, then it is likely that a disproportionate number of engineering majors will be chosen.

Probably the only place where a true random sample can be taken at a university is in the files of the administration building. Having a computer select random ID numbers from a list of all students could assure that each student has an equal chance of being represented in the sample.

When random selection in a large population is desired and subcategories of the population are of interest, a **stratified sample** may be taken. To do this, we select a certain part of the sample from each subgroup of the population. In the previous example, we may want to assure that we include male students from each major, some from day classes, some from night classes, some of each class (freshmen, sophomores, juniors, seniors, graduate students), and some from each ethnic group. The proportion of students from each subgroup of the sample must be the same as the proportion of that subgroup in the entire population. Statistics may then be compiled on the entire sample, or on any subgroup.

A sample cannot accurately represent a population unless it is drawn without bias. **Bias** means that there are factors operating on the sample to make it unrepresentative of the population. These factors are sometimes very subtle, and we may not even be aware of them. But if the sample is totally random and sufficiently large, even unknown factors that may bias the results will be eliminated or distributed within the sample in the same way they are distributed in the population.

Parameters and Statistics

The only way to know the exact characteristics of a population is to measure all of its members. A **parameter** is a characteristic of the entire population. A **statistic** is a characteristic of a sample that is used to estimate the value of the population parameter.

Any estimate of a population parameter based on sample statistics contains some error. The amount of error is never known exactly, but it can be estimated based on the size and variability of the sample. For example, if we measure all males at a university and determine that the average percent body fat in the population is 21%, this value is called a parameter. If we then take a sample and find that the percent fat in the sample is 18%, this value is called a statistic. The difference of 3% between the parameter and the statistic is the result of sampling error. In chapter 6 we will learn how to estimate the amount of error in a sample.

Probability and Hypothesis Testing

Statistics has been called the science of making educated guesses. Unless the entire population is measured, the only statement that can be made about a population based on a sample is an educated guess accompanied by a probability statement. Statistics allow us to make a statement and then cite the odds that it is correct. For example, if we wanted to know the height of 15-year-old females in a city, it might be stated that based on a random sample of 200 females the average population height is estimated to be 5'2" with an error factor of ±2 inches and that the odds that this estimate is correct are 95 to 5. In statistical terms, this is called a 95% level of confidence, or a 5% probability of error. This error factor is usually represented as $p < .05$ (p is almost never exact, so we usually say "the probability of error is less than 5%"). In addition, p < .05 also indicates that the probability that the null hypothesis is true is less than 5%.

Another approach might be to hypothesize (guess or predict) what the average population height is likely to be and then use a sample and a probability statement to test the hypothesis.

Suppose we hypothesize that the average height of the population of 15-year-old females in a city is 5'3". Using the data from the sample of 200 described earlier, we infer that there is a 95% chance that the population average height is 5'2" with an error factor of ±2 inches (chapter 6 describes how to predict population parameters using statistical inferences). This means that the odds are 95 to 5 that the true average height of the population of females is between 5'0" and 5'4". Under these conditions, we accept as true the hypothesis that the height is 5'3" because it lies within the limits of the values estimated from the sample. However, a hypothesis that the average or mean height is 5'5" would be rejected because that value does not fall within the limits of the estimated population value.

The technique of hypothesizing and then testing the hypothesis by experimentation provides answers to questions that cannot be directly answered in any other way. Herein lies the real value of inferential statistics: the ability to make a statement based on research and estimate the probability that the statement is correct.

Theories and Hypotheses

A **theory** is a belief regarding a concept or a series of related concepts. Theories are not necessarily true or false. They are only productive or nonproductive in producing hypotheses. Fruitful theories produce many hypotheses that can be tested by the scientific process of experimentation. A hypothesis is stated more specifically than a theory and can be tested in an experiment. When the experiment is complete, the odds that the hypothesis is correct or incorrect can be stated. We determine these odds by using statistical procedures.

When a theory produces many hypotheses, most or all of which are deemed to be correct (by original and confirming studies), the theory is accepted as true. Usually this process goes through many stages, with revisions of the theory as the process proceeds. The process may take years, decades, or even centuries. Finally the theory is accepted as the most correct interpretation of the data that have been observed. When many or most of the hypotheses produced by a certain theory are rejected or cannot be confirmed by other unbiased scientists, the theory is revised or abandoned.

An example of this process in the field of motor behavior is found in the concept of mental practice, or visualization. For many years, scientists and athletes believed, but could not prove, that the learning of motor skills was best accomplished by a combination of mental and physical processes. Athletes seemed to intuitively know that they could improve performance by visualizing themselves executing the skill. This theory produced many hypotheses, which were tested. One popular hypothesis was that if a person just thought about a physical skill, without ever having performed it physically, his or her performance on the skill would improve.

A few studies in the mid-20th century produced some hints that this may indeed be true. One early researcher in this area (Twinning, 1949) published a study that demonstrated significant differences in physical performance between control subjects and subjects who visualized themselves performing a novel skill. Soon other studies with similar, but modified, hypotheses were adding to the knowledge base.

In the 1960s many studies (Egstrom, 1964; Jones, 1965; Oxendine, 1969; Richardson, 1967; Vincent, 1968) based on visualization theory were published, most of which confirmed the conclusion that visualization groups perform significantly better than control groups, but not as well as physical practice groups. The theory, or general belief, produced testable hypotheses that were supported by experiments. Today, the theory that visualization improves performance of a physical skill is well accepted.

Misuse of Statistics

Advertisements tend to quote statistics freely. We are told that a certain toothpaste is better than another, that "with our magic exerciser, you can lose 10 pounds per week with only 1 minute of exercise per day," or that "8 out of 10 doctors recommend our product." These are examples of statistics that may or may not be true. Such misleading statements usually result from an inadequate definition of terms, the lack of a random sample, or too small a sample; a statement may have no statistical analysis for support or be just an uneducated guess.

One educational institution claimed that 33% of coeds marry their professors. Although this was true at the time for that school (only three women were enrolled in the school and one married a professor), it is an example of an improper generalization from a small sample. It leads the reader to an incorrect conclusion.

Sometimes the word *average* is misleading. For example, it is possible to show that more than half of the population has an annual income lower than the average. If 9 persons make $10,000 per year, and 1 makes $100,000, then 9 out of 10 make less than the average income of $19,000. Perhaps a better description of the income of the population would be the typical income (the 50th percentile). Extreme scores, sometimes called *outliers,* have a disproportionate effect on the statistics. This is somewhat analogous to the joke about the man with his head in the oven and his feet in the freezer who remarked, "On the average, I feel fine."

Statistics is the process of making educated guesses. A statistician attempts to explain or control the random effects that may be operating and to make a probability statement about the remaining real effects. Statistics do not prove a statement. At best, they give us enough information to determine the probability that the statement is true.

We need to be careful when producing or using statistics. We should report only well-documented, valid, reliable, and objective results. And we should be wary of unusual or unique claims, especially if the source of the statistics is likely to benefit financially when we believe the claims.

Summary

This chapter introduced the concepts of measurement, statistics, and research, discussing the essentials of measurement theory and the process of how things are measured.

The interrelationships among measurement principles, statistics, and research design were also discussed. Research design and statistical analysis are always interrelated: Research cannot be conducted without statistical analysis, and proper statistical analysis depends on the quality of the research design.

The process of hypothesis testing and the appropriate use of the research and the null hypothesis were introduced. The use of statistical inference to estimate population parameters from sample statistics is an essential tool of the statistician.

Problems to Solve

1. Define the following terms and state their relationship to one another: (a) measurement, (b) data, (c) statistics, and (d) evaluation.
2. What is the difference between a variable and a constant? Between independent and dependent variables? Give examples of each.
3. List the four classes of data and provide an example of each one. Explain why your example fits the class in which you placed it.
4. Explain how statistical inference is conducted. Define population, sample, random sample, and stratified random sample.
5. What is the difference between a parameter and a statistic?
6. What is the relationship between a theory and a hypothesis? Give an example of how a hypothesis may be used to solve a problem in kinesiology.
7. Give an example of a misuse of statistics that you have observed.
8. List three journals in the general field of kinesiology where research and statistical conclusions are presented.

See appendix C for answers to problems.

Key Words in This Chapter

Measurement
Data
Statistics
Evaluation
Validity
Reliability
Objectivity
Variable
Constant
Continuous variable
Discrete variable
Nominal scale
Frequency data
Ordinal scale
Interval scale
Ratio scale
Research
Historical research
Descriptive research
Experimental design
Hypothesis

Research hypothesis
Null hypothesis
Independent variable
Dependent variable
Predictor variable
Criterion variable
Internal validity
Intervening variable
Instrument error
Investigator error
External validity
Statistical inference
Population
Sample
Random sample
Stratified sample
Bias
Parameter
Statistic
Theory

Chapter 2

Organizing and Displaying Data

The physical education teachers at a middle school decided to test their students on selected fitness items. They used the pull-up test to measure upper body strength and the 1-mile run to measure aerobic capacity. Each instructor tested his or her own classes and recorded the data in the class roll books. Next they decided to place the results for all the students (more than 700) into one database so they could compare the fitness scores across all classes. When the instructors finished entering the information, they found that they had an enormous database that was cumbersome and difficult to read. How should the data be organized?

Organizing Data

In chapter 1, statistics was defined as a mathematical technique by which data are organized, treated, and presented for interpretation and evaluation. Computers are wonderful devices that can organize, analyze, and display data much faster than humans. You should use a computer whenever possible. However, computers are only an extension of the human brain. They will not perform well unless someone enters the data correctly and understands the output. Before a computer will be of value to you, you must know what you want it to do and what to expect in the output. This chapter demonstrates techniques for organizing and displaying raw data into a readable and useful format. It explains how to display the data in tables and how to create graphs so that the data can be more easily interpreted. Once you understand the methods involved, you should use a computer to speed up the process.

The data we collect contain much information of potential value. But before we can derive information from raw data, we must organize the data into a logical, readable, and usable format called a distribution table. Three types of distribution tables are discussed in this chapter: the rank order distribution, the simple frequency distribution, and the grouped frequency distribution.

Rank Order Distribution

A **rank order distribution** is an ordered listing of the data in a single column. It presents a quick view of the spread or variability of the group. It easily identifies the extreme scores, the highest and the lowest. A rank order distribution is used when the number of data points (N) is relatively small (i.e., ≤ 20). If the ordered listing of the data can fit on one page, then the rank order distribution is appropriate.

The **range** (R) is the distance in numerical value from the highest (H) to the lowest (L) score. It is calculated by subtracting the lowest score from the highest score:

$$R = H - L. \tag{2.01}$$

Some statisticians add 1 to this formula so that $R = H - L + 1$. The choice of formula for the range depends on how one looks at the range. If the values at both ends of the range are counted, then the range of 5 to 1 is 5 ($R = 5 - 1 + 1$). But if the range is thought of as the number of steps from L to H, then the range of 5 to 1 is 4 ($R = 5 - 1$). Either position is defensible; in this text, we will follow the first option, equation 2.1.

The range is not easily determined until the raw data have been ordered. Following is a subset of data taken from one of the teacher's roll books in the department described at the beginning of the chapter. The data represent the number of pull-ups performed by 15 eighth-grade boys.

Raw Data: 12, 10, 9, 8, 2, 5, 18, 15, 14, 17, 13, 12, 8, 9, 16.

These data are difficult to interpret because they are not organized. When the data are placed into a rank order distribution, from highest to lowest (see table 2.1), they are easier to interpret. The symbol X represents a variable, in this case pull-ups, and N (the number of scores) = 15.

When the pull-up scores are organized into a rank order distribution, it is easy to see that there is a large difference in ability to perform pull-ups. The spread of the group, or the range, is considerable—from 2 to 18 pull-ups.

Simple Frequency Distribution

When the number of cases being studied is large, it is inconvenient to list them separately because the list would be too long. Larger data sets can be organized into a **simple frequency distribution,** an ordered listing of the variable being studied (X), with a frequency column (f) that indicates the number of cases at each given value of X. (This pattern is referred to as a simple frequency distribution because there is also a more complicated distribution called a *grouped* frequency distribution, which we will consider next.)

Suppose a student majoring in kinesiology wanted to analyze the pull-up scores for all male kinesiology majors at the university ($N = 212$). To avoid a long list, the student could arrange the scores into the simple frequency distribution presented in table 2.2.

Table 2.1   Rank Order Distribution of Pull-Up Scores

X
18
17
16
15
14
13
12
12
10
9
9
8
8
5
2

$N = 15$
$H = 18$
$L = 2$
$R = 18 - 2 = 16$

Table 2.2   Simple Frequency Distribution of Pull-Up Scores

X	f
20	2
19	0
18	3
17	6
16	8
15	10
14	17
13	21
12	25
11	24
10	26
9	19
8	16
7	12
6	10
5	4
4	3
3	2
2	1
1	2
0	1
	212

$$N = 212$$
$$H = 20$$
$$L = 0$$
$$R = 20 - 0 = 20$$

The left column in table 2.2 labeled X represents the variable (pull-up scores), and the right column labeled f represents the number of subjects (in this case, males) who received a given score. The number of subjects measured (N) is represented by the sum of the frequency column (212). Use this type of simple frequency distribution when $N > 20$, and $R \leq 20$. This will allow you to fit all of the data on one sheet of paper.

Grouped Frequency Distribution

It's fairly easy to organize data into a simple frequency distribution if the range of scores is only 20, but researchers often work with variables that produce a range of more than 20 scores. When the physical education teachers looked at the students' aerobic capacity as measured by the mile-run scores ($N = 206$), they found that the lowest score, or fastest time, was 302 seconds and the highest score, or slowest time, was 595 seconds—a range of 293 seconds ($595 - 302 = 293$). With such a range, it is impractical to list all of the scores in one line. The teachers needed a method of grouping the data that would reduce the length of the raw data list.

A **grouped frequency distribution** is an ordered listing of a variable (X) into groups of scores in one column with a listing in a second column, the frequency column (f), of the number of persons who performed in each group of scores. When $N > 20$ and $R > 20$, a grouped frequency distribution should be used.

When setting up a grouped frequency distribution, we must first decide how many groups should be formed. For convenience, we will adopt 15 as the ideal number of groups. Fifteen was chosen because it represents a reasonable number of groups to list on one page of paper, but the decision is an arbitrary one.

When the number of groups becomes as small as 10, the number of scores in each group may become too large. This tends to obscure the data because so many cases are crowded together. Likewise, when there are more than 20 groups, the groups are spread out so far that some groups may lack cases and the list becomes long and cumbersome. For these reasons, it is best to keep the number of groups at about 15, but the number may vary between 10 and 20.

In the example of the mile-run times, groups of 20 scores each (e.g., 300–319) produce 15 groups (see table 2.3).

Table 2.3   Grouped Frequency Distribution: Mile-Run Times in Seconds

X	f
580-599	3
560-579	9
540-559	13
520-539	15
500-519	17
480-499	21
460-479	19
440-459	25
420-439	23
400-419	18
380-399	15
360-379	12
340-359	9
320-339	5
300-319	2
	$N = 206$

The formula for i (the size of a group of scores on a given variable, known as group **interval size**) is

$$i = \text{Range}/15. \qquad (2.02)$$

For the mile-run data, $i = 293/15 = 19.53$, which we round to 20.

Grouping scores is much easier if we can place scores into groups that have whole numbers for their starting and ending points. But depending on the range, i may be either an integer or decimal value. To simplify the process of grouping, we usually round i to the nearest odd integer. An odd interval size always results in a group with a whole number as the midpoint.

The one exception to the odd rounding rule is when i is equal or close to 10 or 20 (as is the case in our example). Because we are more familiar with 10 than with any other number, 10 makes a convenient i size. Even though the midpoint will not be a whole number, i values close to 10 or 20 are usually rounded to 10 or 20 for convenience in creating groups and tallying raw data into the groups. The decision to round to an odd integer, or to 10 or 20 when i is near these numbers, is another arbitrary decision. These choices are based on the nature of the data and on the need for simplicity in tallying the scores into groups.

After i is determined, we next create the groups. When i is a multiple of 5 (or 0.5 if the data are recorded in tenths), the lowest group should include the lowest score in the data, but the lower limit of the lowest group should be a multiple of 5. When i is not a multiple of 5, the lower limit of the lowest group should be equal to the lowest score in the data. Because the determination of i is an arbitrary decision, the rounding-off procedure may be modified to create groups with which it is easy to work.

Because $i = 20$, we want 20 scores per group, and the first group should start with a multiple of 5. The lowest score is 302, so the first group will include all scores between 300 and 319 (see table 2.3). Note that the upper limit of the first group is 319, not 320. The group starts at 300, and the 20th score up from 300 is 319. In a similar manner, the remaining groups are created, with 20 scores in each group. The group creation process is continued up the data scale until a group is created (580–599) that includes the highest score (595).

After the groups are created, tally the scores that fall into each group. The frequency of each group is determined by adding the tally marks. Usually tallying is done on scratch paper; the final result shows only the groups and the frequency, as is shown in table 2.3. Grouping the scores makes it easier to interpret the data. In table 2.3, most values are in the middle groups, and the average is about 450 seconds. Since group frequency tables are used when N is large, a computer can be a great help when dealing with large databases that require frequency distribution tables. Computer software is available that will allow the user to enter data and organize it in various ways.

Some information is lost during grouping. For example, once the scores are grouped, it is impossible to tell exactly which score a given person received. We know that 25 subjects scored between 440 and 459 seconds, but we do not know how the scores are distributed within the group. Did all 25 subjects score 450? Did anyone score 440, 445, or 455? We assume that the scores in a group are equally distributed within that group and that the mean score falls at the midpoint of that group. Using whole numbers for the midpoints makes working with the data easier.

Real and Apparent Limits of Groups

The process just described for grouping data works well for discrete data in which decimal values are not involved. However, a problem arises when we use continuous data and measure scores in fractions of whole numbers. Consider the grouped frequency distribution representing time in seconds to swim 100 meters shown in table 2.4.

Table 2.4   Grouped Frequency Distribution: Times in Seconds for 100-Meter Swim

X	f
115-119	1
110-114	2
105-109	3
100-104	7
95-99	10
90-94	15
85-89	17
80-84	11
75-79	5
70-74	4
65-69	2
60-64	2
	$N = 79$

Swim times are usually measured in tenths, or even hundredths, of a second. The grouped frequency distribution is set up by integers, but the raw data include decimals. A student who swam the distance in 86.4 seconds would obviously be included in the 85–89 group, but into which group would we place the student who scored 84.7 seconds? Because time scores can be recorded to any degree of accuracy, we must make some provision to accommodate these in-between scores. We accomplish this by establishing the real limits of the groups.

The **apparent limits** of each group are the integer values listed for each group in the grouped frequency distribution. There is a one-point gap between the apparent end limit of one group and the beginning of the next. The **real limits** of each group are the assumed upper and lower score values in each group, created to the degree of accuracy required by the data; the real limits define the true upper and lower limits of the group. To establish the real limits, the gap between the apparent limits is equally divided; values in the upper half belong to the group above, and values in the lower half belong to the group below.

The lower real limit of the 80–84 group is 79.5, and the upper real limit is 84.49999 . . . . Usually the upper real limit is carried out to one decimal place more than the data requires (i.e., if the data is in tenths, set the limit at hundredths).

Thus, to establish the real limits of the 100-meter swim groups, we raise the highest score in the group by 0.49 and lower the lowest score in the group by 0.5. Once this is done, it is easy to determine that the swimmer whose time was 84.7 seconds should be included in the 85–89 group because the lower real limit of that group is 84.5. The real limits are not listed in the grouped frequency distribution, so we need to keep those values in mind and classify the data into groups according to the real, not the apparent, limits.

Displaying Data

Tables present data in a row-and-column format. Tables are useful for organizing and determining the limits of the data. Graphs are another useful way to present information about data. Graphs are visual representations of data that provide us with insight into the data that may not be easily observed in a table.

Graphs

A **graph** is a figurative, or visual, representation of data. Graphs are helpful for comparing two or more sets of data or for showing trends. All graphs should include at least three characteristics: (a) a title, (b) a label on the X-axis (the abscissa), and (c) a label on the Y-axis (the ordinate). Other explanatory notes and information may be included on the graph.

It is a general practice on statistical graphs to place the frequency on the ordinate (the vertical or Y-coordinate) and the scores on the abscissa (the horizontal or X-coordinate), but this practice is not mandatory. Reversing this process would rotate the graph 90 degrees, but the data would not be changed. On the X-axis, scores move from left to right as the numeric values move from low to high. But good scores are not always high scores—for example, faster runners have lower time scores. On the Y-axis, frequencies move from bottom to top as the numeric values move from low to high.

Histogram

Probably the most common graph is the **histogram,** or **bar graph.** It is usually constructed from a grouped frequency distribution, but it may also originate from a simple frequency distribution. The bars on the X-axis represent the groups, and the height of each bar is determined by the frequency of that group as plotted on the Y-axis. The data from table 2.4 (100-meter swim times) are shown in figure 2.1 in a histogram. The apparent limits of each group are plotted on the X-axis.

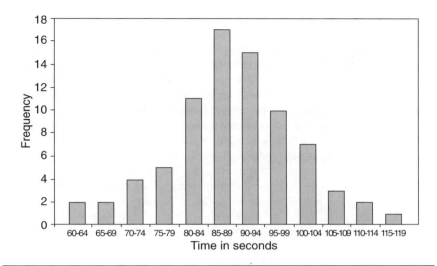

Figure 2.1   Histogram of 100-meter swim.

Another graph that may be constructed from frequency distribution data, called a **frequency polygon,** is a line graph of the scores plotted against the frequency. It is usually formed from simple frequency distribution data by plotting the ordered scores on the X-axis against frequency on the Y-axis. A frequency polygon of the data from table 2.2 on pull-up scores is shown in figure 2.2.

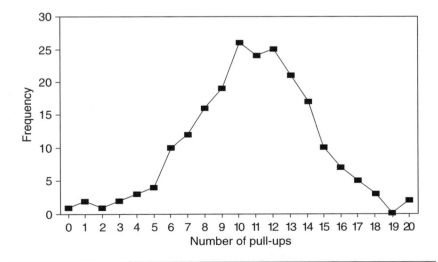

Figure 2.2   Frequency polygon for pull-up scores.

Cumulative Frequency Graph

A third graph often used by statisticians is called a **cumulative frequency graph,** or cumulative frequency curve. It is a line graph of ordered scores on the X-axis plotted against the number of subjects who scored at or below a given score on the Y-axis. An exercise physiologist, studying the effects of resistance exercises, measured upper body strength on a group of subjects and counted the number of arm dips subjects could perform on the parallel bars. The scientist arranged the resulting data into a grouped frequency distribution and constructed a grouped frequency table that included a cumulative frequency column (see table 2.5).

The cumulative frequency column tells us the status of each subject compared to the total group. It is calculated by determining the number of subjects whose score is at or below that of each group. To accomplish this, we add the frequency of a given group to the frequencies of all the groups below it. To determine the cumulative frequency of the 10–12 group, we add 6 + 10 + 15 + 18 = 49.

The cumulative frequency graph is plotted with scores on the X-axis and cumulative frequency on the Y-axis. Because the data in table 2.5 are discrete, the upper apparent limit of each group is plotted because that value represents all subjects who scored at or below that group. For continuous data, the upper real limits would be plotted to represent the maximum score that could be obtained by anyone in that group or in the groups below. Figure 2.3 is an example of a cumulative frequency graph for the parallel bar dip test data from table 2.5.

Table 2.5   Upper Body Strength as Measured by Parallel Bar Dip Test

X	f	Cum. f
34-36	2	130
31-33	4	128
28-30	5	124
25-27	7	119
22-24	9	112
19-21	15	103
16-18	19	88
13-15	20	69
10-12	18	49
7-9	15	31
4-6	10	16
1-3	6	6
	$N = 130$	

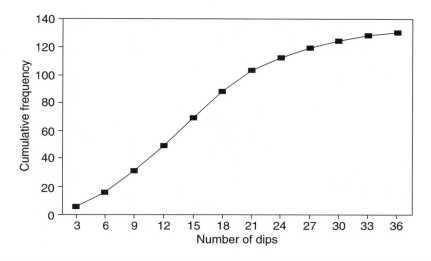

Figure 2.3 Cumulative frequency of dips on parallel bars.

Curves

A **curve** is the line that results when scores (*X*) are plotted against frequency (*Y*) on a graph. The shape of the curve depends on the distribution of the data. A curve presents a visual picture that permits us to see trends in the data that we may not easily observe when looking at a table. The following curves are used most often in statistics.

Normal Curve

The most widely known curve in statistics is the **normal curve.** This uniquely shaped curve, which was first described by mathematician Karl Gauss (1777–1855), is sometimes referred to as a Gaussian curve, or a *bell-shaped curve.* The normal distribution forms the basis of all statistical reference.

> Nature generally behaves according to rule. Karl Friedrich Gauss discovered that fact and formulated his discovery in a mathematical expression of normal distribution . . . and this curve has ever since become the sine qua non of the statistician. (Leedy, 1980, p. 25)

A normal curve is characterized by symmetrical distribution of data about the center of the curve in a special manner. The mean (average), the median (50th percentile), and the mode (score with the highest frequency) are all located at the middle of the curve. The frequency of scores declines in a predictable manner as the scores deviate farther and farther from the center of the curve.

All normal curves are bilaterally symmetrical and are usually shaped like a bell, but not all symmetrical curves are normal. When data are identified as normal, the special characteristics of the normal curve may be used to make statements about the distribution of the scores. Many of the variables measured in kinesiology are normally distributed, so kinesiology researchers need to understand the normal curve and how it is used. Chapter 6 discusses the special characteristics of the normal curve in detail.

A typical normal curve is presented in figure 2.4. Notice that the two ends, or tails, of the curve are symmetrical and that they represent the scores at the low and high extremes of the scale. When scores approach the extreme values on the data scale, the frequency declines. This is demonstrated by the data represented in table 2.4. Most subjects score in the middle range.

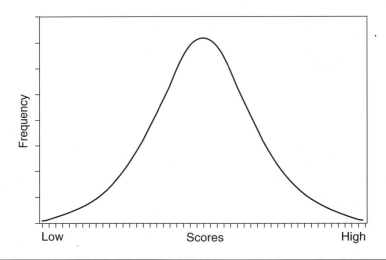

Figure 2.4   Normal curve.

Mesokurtic, Leptokurtic, and Platykurtic Curves

When most scores fall in the midrange and the frequency of the scores tapers off symmetrically toward the tails, the familiar bell-shaped **mesokurtic** (*meso* meaning middle and *kurtic* meaning curve) curve results. But if the range of the group is limited and the extreme scores are very close to the middle, with most subjects scoring near the mean, the curve is called **leptokurtic,** or peaked.

The opposite of the leptokurtic curve, which results from a very wide range of scores with low frequencies in the midrange, is called **platykurtic,** or flat. The differences among mesokurtic, platykurtic, and leptokurtic curves are shown in figure 2.5.

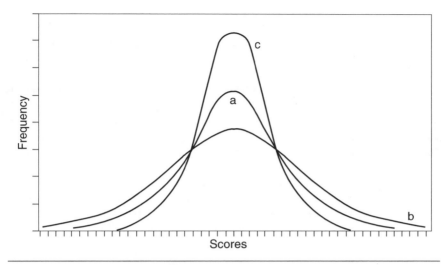

Figure 2.5   (a) Mesokurtic (bell-shaped), (b) platykurtic, and (c) leptokurtic curves.

Bimodal Curves

The **mode** is the score with the highest frequency. On the normal curve, a single mode is always in the middle, but some distributions of data have two or more modes and hence are called **bimodal,** or multimodal. If one mode is higher than the other, the modes are referred to as the major and minor modes. When such data distributions are plotted, they have two or more humps representing the clustering of scores. Bimodal curves are not normal curves. A bimodal curve is shown in figure 2.6.

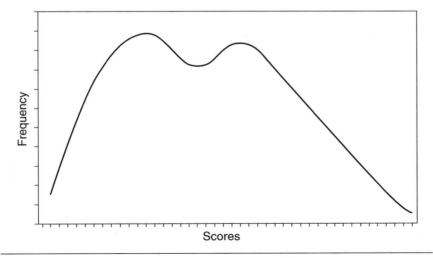

Figure 2.6   Bimodal curve.

Skewed Curves

Sometimes the data result in a curve that is not normal; that is, the tails of the curve are not symmetrical. When a disproportionate number of the subjects score toward one end of the scale, the curve is **skewed.** The data in table 2.5 for parallel bar dips show that a larger number of the subjects scored at the bottom of the scale than at the top. A few stronger subjects raised the average by performing 30 or more dips. When the data from table 2.5 are plotted, as in figure 2.7, the hump or mode of the curve is pushed to the left, and the tail on the right is longer than the tail on the left. The curve has a positive skew because the long tail points in a positive direction on the abscissa.

If most subjects score high on a test and only a few do poorly, then the mode is to the right, and the left tail is longer than the right tail, as shown in figure 2.8. This is called a negative skew, because the long tail points in the negative direction on the abscissa. Chapter 6 presents a method for calculating the amount of skewness in a data set.

Other Shapes

Other curve shapes are possible, but most of them are relatively rare. **Rectangular curves** occur when the frequency for each of the scores in the middle of the data set is the same. A **U-shaped curve** is the result of a high frequency of values at the extremes of the scale and a low frequency in the middle.

A **J-curve** results when frequency is high at one end of the scale, decreases rapidly, and then flattens and reduces to almost zero at the other end of the scale. This

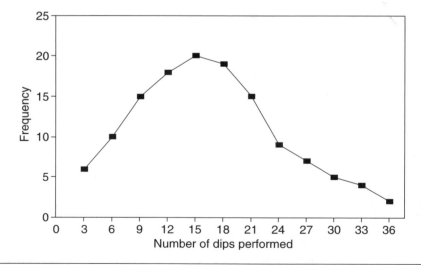

Figure 2.7   Positively skewed curve for dips on parallel bars.

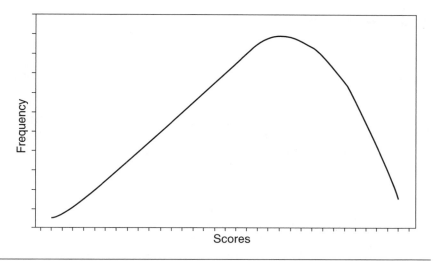

Figure 2.8 Curve with a negative skew.

curve is different from a straight line, in which frequency decreases uniformly from one end of the scale to the other. J-curves can be positive or negative in direction, depending on the orientation of the lower tail of the curve. If the tail points to the positive end of the X-axis, the curve is positive. J-curves may also be inverted.

Summary

The purpose of organizing and displaying data is to produce order and parsimony from raw data. After you understand the principles explained in this chapter, you should use a computer to organize and display your data. Organizing and display-ing data is the first step in a statistical analysis. Organizing data requires that we present them as a distribution of scores. A rank order distribution can be used when the sample size is ≤ 20. When the range is limited to 20 or fewer values, but the number of scores is large ($N > 20$), a simple frequency distribution is used. When both the range and the number of scores are large, the data should be ar-ranged into a grouped frequency distribution.

A graph presents a visual picture of data and reveals trends that may not be obvious in a table. The curve of the graphed data is especially meaningful because it shows how the data are distributed. Many of the variables measured in kinesiol-ogy are normally distributed. The normal curve has special characteristics that allow us to make statements about the distribution of scores. Chapter 6 will discuss the normal curve in more detail.

Problems to Solve

The following scores are from a 4 × 10-yard shuttle-run test for a group of middle school students. The scores are measured in seconds to the nearest tenth of a second.

Raw Data: 12.1 12.9 13.1 13.6 15.4 12.2 12.3 11.9 11.8 10.4 10.8 13.0 9.3 11.4
13.8 11.8 10.9 13.4 12.7 10.8 14.8 15.9 11.0 11.7 14.2 12.9 10.3 9.8
12.0 13.1 14.7 11.8 10.5 11.7 9.6 12.1 14.4 15.3 12.1 10.3 11.2 11.4
10.1 11.1 13.4 12.6 9.1 13.0 12.6 12.7 11.9 10.4 11.6 12.5 13.2 13.3
11.5 10.6 12.3 12.9 13.5 12.8 11.6 12.8 12.5 12.0 12.4 14.6 12.8 12.4
13.6 11.7

1. Determine the highest and lowest scores, and calculate the range.
2. Determine an appropriate interval size (i) for grouping data. Hint: Remember that you are working with data accurate to the nearest tenth of a second.
3. Establish the groups and tally the number of scores per group. Develop a frequency column and a cumulative frequency column. Set the lower limit of the lowest group as a multiple of 0.5.
4. What are the real limits of group 12.0–12.4?
5. Draw a histogram from the data.
6. Draw a frequency polygon from the data. Are the data normal or skewed?
7. What is likely to happen to the frequency polygon if more cases are added to the data?
8. Draw a cumulative frequency graph from the data.

See appendix C for answers to problems.

Key Words in This Chapter

Rank order distribution

Range

Simple frequency distribution

Grouped frequency distribution

Interval size

Apparent limits

Real limits

Graph

Histogram

Frequency polygon

Cumulative frequency graph

Curve

Normal curve

Mesokurtic

Leptokurtic

Platykurtic

Mode

Bimodal

Skewed

Rectangular curve

U-shaped curve

J-curve

Chapter 3

Percentiles

When basketball announcers compare the performances of various players or teams, they usually use percentages rather than raw data. We see the shooting percentages of certain players compared on free throws, 2-point shots, and 3-point shots. Percentages are used because everyone understands them, and they allow immediate comparisons of two or more players. Raw data would make it hard to tell who is a better free throw shooter: player A, who made 46 of 62 attempts, or player B, who made 56 of 78. But when the data are converted to percentages (74% for player A, 72% for player B), comparisons of data can easily be made.

Percentiles are probably the most common statistical tool in use today. Educators and scientists find percentiles helpful in interpreting data for the lay public. This chapter presents methods of calculating percentiles by hand. Although researchers use computers to make calculations on more complicated data or on databases with large numbers of scores, it's best to learn the methods using simple data and paper and pencil.

Cent is the Latin root word for one hundred. The word *percent* means by the hundred. When we use percents, we are comparing a raw score with a conceptual scale of 100 points. Percents provide a quick reference for any raw score in relation to the rest of the scores. A percentile represents the fraction (in hundredths) of the ordered scores that are equal to or fall below a given raw score. In statistics, a **percentile** is defined as a point or position on a continuous scale of 100 theoretical divisions such that a certain fraction of the population of raw scores lies at or below that point.

A score at the 75th percentile is equal to or surpasses three-fourths of the other scores in the raw data set. The 33rd percentile is equal to or better than about one-third of the scores but is surpassed by two-thirds of the scores. A student's percentile score on a test indicates how the student's score compares with the other raw scores.

Percentiles are standard scores. A **standard score** is a score that is derived from raw data and has a known basis for comparison. In percents, the center is 50% and the range is 0% to 100%. A middle-aged male may be able to consume 40 milliliters of oxygen per kilogram of body weight per minute. This is the raw score. Without more information, the score is difficult to evaluate. But if we compare this value to other values in the population of all males of like age and calculate a percentile score of 65, then we know that the man's oxygen consumption is equal to or better than that of 65% of people in that population. Raw scores are measured values. Standard scores, which are derived from raw scores, provide more information than do raw scores.

Standard scores allow us to (a) evaluate raw scores and (b) compare two sets of data that are based on different units of measurement. Which is the better score, 150 feet on the softball throw or 30 sit-ups in a minute? With only this information, it is impossible to tell. We must ask, What was the range? What was the average?

But when we compare percentile scores of 57 on the softball throw versus 35 on the sit-up test, the answer is clear. Not only are the relative values of the scores apparent, but we also know the middle score (50) and the range (0–100). The student who received these scores is better on the softball throw than on sit-ups in comparison to the other students in the class.

The conversion of raw scores to standard scores, of which percentiles are just one example, is a common technique in statistics. It is most useful in evaluating data. (Chapter 6 discusses other standard scores.)

Percentiles may present a problem of interpretation when we consider the extreme ends of the scale for data sets with large numbers of scores. The problem results from imposing a percentile scale on interval or ratio data.

For example, figure 3.1 shows a graph of the number of curl-ups performed in 1 minute by 1,000 seventh-grade boys. Both the number of curl-ups (raw scores) and the percentile divisions have been plotted. Note that a score of 30 curl-ups is equal to the 50th percentile.

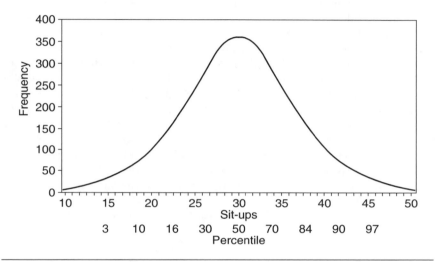

Figure 3.1   Curl-ups per minute. $N = 1,000$.

If a boy scores 30 curl-ups on the first test and raises his raw score to 35 on a second trial, his percentile score would raise from 50 to about 75, or 25 percentage points. However, if he were already at the upper end of the scale and made the same 5 curl-up improvement from 45 to 50, his percentile would only increase about 3 points, from 97 to near 100. This phenomenon is sometimes called the **ceiling effect.**

Teachers and coaches must account for this when they grade on improvement. It is much easier to lower the time it takes to run the 100-meter dash from 12.0 seconds to 11.5 seconds than it is to lower it from 10.5 seconds to 10.0 seconds. An improvement of 0.5 seconds by a runner at the higher level of performance (i.e., lower-time scores) represents more effort and achievement than does the same time improvement in a runner at the middle of the scale.

Learning curves typically start slow, accelerate in the middle, and plateau as we approach high-level performance. The ceiling effect is demonstrated by the plateau in the curve. It is more difficult to improve at the top of the learning curve than at the beginning or in the middle.

This difficulty in interpretation should not deter us from using percentiles, but we must recognize the ceiling effect when considering scores that represent high levels of performance, particularly if improvement is the basis for evaluation.

Common Percentile Divisions

All measurements contain some error. When the error is relatively large, the subject's true score is only approximated. In these cases, it would be inappropriate to report an exact percentile rank because the raw score is only an estimate of the true score. Reporting a range of scores or a range of percentiles within which the subject's true score probably lies is a common practice.

The two most common ranges of percentile scores are the **quartile** range and the **decile** range. Occasionally, scores are reported in **quintiles.** In the quartile range, the percentile scale is divided into four equal parts, or quartiles. The 1st quartile extends from 0 to the 25th percentile (Q_1). The 2nd quartile ranges from Q_1 to the 50th percentile (Q_2). The 3rd quartile extends from Q_2 to Q_3 (the 75th percentile) and the 4th, or highest, quartile reaches from Q_3 to Q_4 (the 100th percentile). In the quintile range, five divisions of the percentile scale are made.

Deciles follow the same format except the scale is divided into ten parts, each of which has a range of 10 points. The 1st decile ranges from 0 to D_1, the 2nd from D_1 to D_2, and so forth; the 10th decile extends from D_9 to D_{10} (the 100th percentile). Figure 3.2 shows the quartile and decile divisions of the percentile scale graphically.

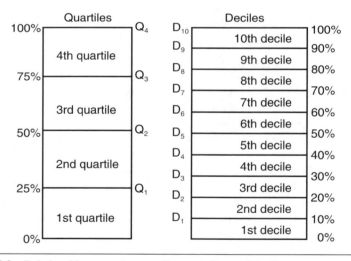

Figure 3.2   Relationships among percentiles, quartiles, and deciles.

Sometimes test scores are reported in quartile or decile ranks. If a student ranks in the 3rd quartile, his or her score lies somewhere between the 50th and the 75th percentiles. Likewise, a decile rank of 4 means the score lies between the 30th and the 40th percentiles. This type of reporting is used if the teacher or scientist is aware of errors in testing or if the test is such that only an estimate of the true score can be obtained. Test batteries that report scores on several aspects of a student's ability, so as to form a profile, or total picture, of the student's talents, are commonly reported in quartile or decile ranks.

Calculations Using Percentiles

Converting raw scores into percentiles or determining the raw score that represents a given percentile are easy processes with rank order data. We need to be able to apply these processes to data that have been organized into simple and grouped frequency distributions. In this section we first discuss how to do these processes with rank order distributions. Then we apply the concepts to simple and grouped frequency distributions.

Rank Order Distributions

A basketball coach had 15 players on a high school team complete a free throw test. Each player took 10 shots from the free throw line, and the scores were recorded. Table 3.1 presents the data in a rank order distribution. With scores such as these on a group this small, it may not be meaningful to calculate percentile scores. But the example is presented here to demonstrate the basic concept of percentile calculation.

Table 3.1   Rank Order Distribution: Basketball Free Throws Completed in 10 Attempts

X
10
9
9
7
7
7
6
5
5
5
4
4
3
3
1
$N = 15$

Determining the Percentile From the Score

What is the percentile rank for a raw score of 6 baskets? Remember, a percentile is a point on a continuous scale of 100 theoretical divisions such that a certain fraction of the population of raw scores lie at or below that point. To calculate the percentile for a score of 6 baskets, we need to determine what fraction of scores falls at or below 6 baskets.

Percentiles are not based on the value of the individual scores but on the order of the scores. In table 3.1, the value of the bottom score could be changed to 2 without affecting the percentile divisions, because it would still be in the same order in the group. The question we must ask is, How *many* scores fall *at or below* 6 in the ordered list of scores?

Counting from the score of 6 down, we note that 9 scores fall at or below 6. There are 15 scores, so the person who scored 6 baskets is 9/15 of the way from the bottom of the scale to the top. This fraction is first converted into a decimal (9/15 = .60), and then the decimal is multiplied by 100 to convert it into a percent (.60 × 100 = 60%). Thus a score of 6 on this test falls 60% of the way from the bottom of the scale to the top and is classified as the 60th percentile.

Any single score could be converted in a similar manner, but several scores were obtained by more than one person. For example, three persons each received a score of 5. What is their percentile rank? Using the method described above, we could calculate that the first score of 5, the sixth score from the bottom, represents the 40th percentile (6/15 = .40, .40 × 100 = 40%). But the top score of 5 is eight scores from the bottom and represents approximately the 53rd percentile (8/15 × 100 = 53.3%).

Do we then conclude that the persons who each made 5 baskets scored between the 40th and the 53rd percentile ranks? No. Because we define percentile as a fraction at or below a given score, we conclude that all are equal to or below a score of 5. And because they all performed equally, they should all receive the same percentile as a fraction of scores *at or below* a given score. Therefore all three have a percentile score of 53.3.

Determining the Score From the Percentile

The most common computation in working with percentiles is calculating the percentile from the raw score, but sometimes the opposite calculation is required. If a coach determines that the 60th percentile is the cutoff point on a test for selecting athletes for a varsity team, the coach needs to know what raw score is represented by the 60th percentile.

Frequently, grades or class divisions are made on the basis of percentiles. A class may be divided into thirds, with the top third receiving one practice schedule, the middle third another, and the bottom third still another.

To determine which students should be included in each group, we must establish the raw scores equivalent to the percentile points. The technique of determining the raw score that matches a given percentile point is the opposite of that for finding the percentile from the raw score.

For example, the basketball coach who collected the free throw data (see table 3.1) can take the top two-thirds of the team to away games. If the coach uses the free throw score as the criterion for determining who makes the traveling squad, how many baskets must a player make to qualify?

The coach must ask, Which score defines the bottom third of the team? All players with scores above that point are in the top two-thirds and qualify for the traveling team. This is a simple problem with 15 players because it is easy to determine that the top 10 players represent the upper two-thirds of the team. But by thinking through the process with simple data we learn the concepts that may be applied to more difficult data.

The percentile equivalent to 1/3 is found by converting the fraction to a decimal (1/3 = .333) and multiplying this decimal equivalent by the total number of subjects in the group (.333 × 15 = 5). This value (5) is the number of scores from the bottom, not the raw score value of 5 baskets made.

To determine the raw score that is equivalent to a given percentile (P), convert the percentile to a decimal, multiply the decimal equivalent by the number of scores in the group (N), and count that many scores from the bottom up. Counting up five scores from the bottom, we find that a score of 4 free throws represents a percentile score of 33.3. Any player who made 4 free throws or less is included in the bottom third of the group, and any player scoring higher than 4 free throws is included in the top two-thirds of the group. So 5 free throws is the criterion for making the traveling squad.

With small values of N and discrete data, it may be necessary to find the score closest to a given percentile if none of the raw scores falls exactly at that point. If the product of $P \times N$ is not an integer, round off to the nearest integer to determine the count from the bottom. In table 3.1, the 50th percentile is 7.5 scores from the bottom (0.50 × 15 = 7.5), but there are no half scores. So we round 7.5 to 8 and count 8 scores from the bottom. The 8th score up is 5. Therefore 5 is the closest score to the 50th percentile.

Simple Frequency Distributions

When more scores are available, but the range is still small ($N \geq 20$ and $R \leq 20$), the scores are usually grouped into a simple frequency distribution. Suppose the basketball coach in the previous example gave the free throw test to 60 students in a physical education class. The data are presented in a simple frequency distribution in table 3.2.

Determining the Percentile From the Score

What is the percentile rank of a student who made 7 baskets? To answer this question, we need to compute the fraction of scores that are equal to or less than 7. We do so by adding the numbers in the frequency column (f) from the score (X) of 7 down to the bottom.

Table 3.2   Simple Frequency Distribution: Basketball Free Throws Completed in 10 Attempts

X	f	Cum. f
10	2	60
9	6	58
8	9	52
7	12	43
6	15	31
5	8	16
4	4	8
3	2	4
2	1	2
1	1	1
	$N = 60$	

If several percentile calculations are to be performed, it is helpful to create a cumulative frequency column. The cumulative frequency column in table 3.2 indicates that 43 persons scored 7 baskets or less. Sixty persons took the test, so 43/60 of the students scored 7 or less.

Converted to decimals, 43/60 = .716, and .716 × 100 = 71.6. Therefore a person who scored 7 baskets ranks equal to or better than about 72% of those who took the test.

Determining the Score From the Percentile

In table 3.2, what score is equivalent to the 75th percentile? The concept of determining the score equivalent to a given percentile from a simple frequency distribution is the same as it was for a rank order distribution. First we establish the number of free throws at or below which 75% of the group scored. This value is determined as follows: .75 × 60 = 45 students.

The cumulative frequency column indicates that 43 students scored 7 or less, and 52 students scored 8 or less. The 9 students who each scored 8 baskets represent the 44th through the 52nd students in the ordered distribution. Therefore, the 45th student is one of these 9. We cannot separate the 9 students (they are all equal in ability), so the 75th percentile is determined to be 8 baskets. Because 9 students tied at 8 baskets, the bottom three-fourths of the group is *approximated* to be those who scored 8 or less, and the top fourth of the group is *approximated* to be those who scored 9 or more.

If the product of the percentile (P) in decimal value times N is not an integer, we would round the product to the nearest integer and find that value in the cumulative frequency column. For example, if the 14th percentile is desired (.14 × 60 = 8.4), we would find where 8 falls in the cumulative frequency column, then read 4 as the number of baskets closest to the 14th percentile.

Grouped Frequency Distributions

Data accumulated from a test in which a wide range of scores are available ($R > 20$) and many persons participated ($N > 20$) may be presented in a grouped frequency distribution. As an example, we will consider the results of a test of softball throw for distance given to 115 students. Table 3.3 presents the raw data in a grouped frequency distribution.

Determining the Percentile From the Score

What is the percentile rank of a student who threw the softball 146 feet? This problem is approached in the same manner as the simple frequency distribution problem except that, because the data are grouped, we do not know where within an interval a given score lies.

Table 3.3 indicates that seven people threw the ball between 140 and 149 feet. We do not know how many of the seven people threw farther than 146 feet and how many threw less, so we must assume that all seven people are equally distributed between the real limits of the interval (139.50–149.49). Real limits are used so that percentiles for both discrete and continuous data may be calculated.

Table 3.3   Grouped Frequency Distribution: Softball Throw for Distance, in Feet

X	f	Cum. f
220-229	2	115
210-219	5	113
200-209	7	108
190-199	7	101
180-189	9	94
170-179	12	85
160-169	10	73
150-159	15	63
140-149	7	48
130-139	9	41
120-129	10	32
110-119	7	22
100-109	6	15
90-99	5	9
80-89	4	4
	$N = 115$	

This problem is now easily solved by applying the following equation:

$$P = \frac{\left(\dfrac{X - L}{i}\right)f + C}{N}, \tag{3.01}$$

where P = percentile, X = raw score, L = lower real limit of interval in which raw score falls, i = size of the interval, f = frequency of the interval in which the raw score falls, C = cumulative frequency of the interval immediately below the one in which the raw score falls, and N = total number of cases.

When equation 3.01 is applied to the raw score of 146, the following results are obtained:

$$P = \frac{\left(\dfrac{146 - 139.5}{10}\right)7 + 41}{115} = .396 \text{ or } 39.6\%.$$

Therefore, a throw of 146 feet is approximately equal to the 40th percentile.

Determining the Score From the Percentile

Using the data in table 3.3, the teacher may desire to determine the middle score, or 50th percentile. How far would a student have to throw the softball to be considered in the top half of the class? To solve this problem, we must first multiply $P \times N$ to determine the interval in which the score falls. With 115 students in the class, the score representing the 50th percentile falls exactly halfway up the cumulative frequency column ($.50 \times 115 = 57.5$). This places the middle score somewhere in the 150–159 interval.

Equation 3.01 may be solved algebraically for X, to obtain the following equation:

$$X = \left(\frac{PN - C}{f}\right)i + L. \tag{3.02}$$

The symbolic notation is the same as in equation 3.01. The problem previously discussed is now easily solved:

$$X = \left(\frac{.50 \times 115 - 48}{15}\right)10 + 149.5 = 155.83 \text{ ft.}$$

The 50th percentile is approximated by a throw of 156 feet.

Summary

Converting raw scores to standard scores is a common technique in statistics. Standard scores let us evaluate raw scores and compare two sets of data that are based on different units of measurement. Because of this, standard scores provide information about data that is not available from raw scores alone. The most common standard scores, percentile scores, are frequently used in kinesiology. Percentile scores can be determined from rank order data, simple frequency data, and grouped data. Percentiles can also easily be converted into raw scores.

Problems to Solve

A test of tennis serve accuracy resulted in the following data:

X
15
12
12
10
9
8
8
7
5
4
4

1. What is the percentile rank of a person who scored (a) 7; (b) 10?
2. What is the nearest score to (a) the 70th percentile; (b) the 45th percentile?

The same test of tennis serve accuracy was administered to three college classes. The scores were arranged into a simple frequency distribution:

X	f	Cum. f
18	1	70
17	2	69
16	3	67
15	3	64
14	5	61
13	5	56
12	6	51
11	8	45
10	10	37
9	7	27
8	8	20
7	5	12
6	3	7
5	3	4
4	1	1
	$N = 70$	

3. What is the percentile of a person who scored (a) 16; (b) 5?
4. Which score most closely represents the (a) 47th percentile; (b) the 80th percentile?

A teacher tested her students to find out how many sit-ups they could do in a minute. She arranged the scores into a grouped frequency distribution.

X	f	Cum. f
70-74	2	83
65-69	4	81
60-64	7	77
55-59	9	70
50-54	10	61
45-49	12	51
40-44	13	39
35-39	11	26
30-34	7	15
25-29	5	8
20-24	3	3

$$N = 83$$

5. What are the percentile ranks of students who scored (a) 32, (b) 39, and (c) 55 sit-ups?
6. What scores best represent (a) Q_1, (b) Q_2, and (c) Q_3?

See appendix C for answers to problems.

Key Words in This Chapter

Percentile

Standard score

Ceiling effect

Quartile

Decile

Quintile

Chapter 4

Measures of Central Tendency

Following the volleyball unit for his physical education class, Mr. Sanchez, an instructor at Sunnyville High School, administered a volleyball-serving test to measure students' serving skills. Each student was given 20 chances to perform an overhand serve into an empty court. Students received 1 point for every serve that landed in. After the group had completed the test, one student asked, "Mr. Sanchez, what was the average on this test?" The student was really looking to learn the meaning of her score—to interpret any single raw score, the central tendency of the data set from which the score is taken must be known.

Measures of **central tendency** are values that describe the middle, or central, characteristics of a set of data. Some terms that may be used to describe the center of a group of scores are most common score, typical score, and average score. These measures provide important information, which allows us to calculate the relationship of a given score to the middle scores of a data set. The three statistical measures of central tendency are the mode, the median, and the mean.

The Mode

The **mode** is the score that occurs most frequently. There is no formula to calculate it; it is found by inspection. In a rank order listing of scores, the mode can be determined by scanning the list of all the scores to determine which is the most frequent. In a simple or grouped frequency distribution, the score with the highest value in the frequency column is the mode.

In a simple frequency distribution, the mode is the score with the highest frequency, but in a grouped frequency distribution, the mode is considered to be the midpoint of the group with the greatest frequency. A distribution can, of course, have more than one mode. If two or more scores or groups have the same frequency, then the variable is said to be bi- or multimodal.

The ease with which it can be determined is one advantage of the mode. It gives a quick estimate of the center of the group, and when the distribution is normal or nearly normal, this estimate is a fair description of the central tendency of the data. The mode is also the only measure of central tendency that can be used with data on an ordinal scale.

The mode also has some disadvantages:

- It is unstable; it may change depending on the methods used for grouping.
- It is a **terminal statistic;** that is, it does not give information that can be used for further calculations.
- It completely disregards the extreme scores—it does not reflect how many there are, their values, or how far they are from the center of the group.

The Median

The **median** is the score associated with the 50th percentile. It can be determined by using the methods of calculating percentiles that were presented in chapter 3. In this chapter, we discuss the median's importance as a measure of central tendency.

The median is the middle score in that it occurs in the middle of the list of scores; it divides the data set in half. The median is also the typical score because it is the single score that best represents the majority of the other values.

In a rank order distribution, when N is odd, the median is the middle score in the range. When N is even, the median falls between two scores. Consider the two examples listed below:

A	B
9	19
8	18
6 ← median	17
4	← median
1	16
	13
	12

Note: When N is even and the median is between two scores, computer programs usually select the higher of the two scores. In example B, the median would be reported as 17. In example A, the median is 6, the middle score. In example B, there is no single score that represents the exact middle.

The calculation of the median does not take into consideration the value of any of the scores. It is based only on the number of scores and their rank order. For this reason, the median is appropriately used on ordinal data and on data that are highly skewed.

An extreme score that is radically different from the other scores in the data set does not affect the median. Consider the following example: Nine people each perform 40 sit-ups, and one does 1,000. The median score for the group is 40, but the average (mean) is 136. The median would still be 40 even if the highest score were 2,000. It is easy to see why the median is called typical of the group. The median is more representative of the majority of scores than is the average when there are radically extreme scores.

The median's quality of not being affected by the size of extreme scores is both an advantage and a disadvantage. Because the median does not consider the size of scores, but only how many there are, it neglects some important information provided by the data—namely, the value of the extreme scores.

The Mean

The **mean** is the statistical name for the arithmetic average. It is represented by a point on a continuous scale and may be expressed as an integer or a decimal value. The mean is the most commonly used measure of central tendency. Its calculation considers both the number of scores and their values. It gives weight to each score according to its relative distance from other scores in the data set.

Because of this feature, the mean is the most sensitive of all the central measures. Slight changes in any of the scores in the group may not affect the mode or the median, but they always change the mean. The chief advantage of the mean is that it considers all information about the data and provides a basis for many additional calculations that will yield still more information.

These characteristics make it the most appropriate measure of central tendency for ratio and interval data. It is not proper to calculate the mean on ordinal data, because distances between scores are not known in ordinal data. (The median would be more fitting in this case. Calculation of the mean requires more information than is provided by ordinal data, whereas the median considers exactly the type of information given by ordinal data—order of scores, but not relative distance between scores.)

The mean's sensitivity may also be a disadvantage. When one or more scores, called **outliers,** are considerably higher or lower than the other scores, the mean is pulled toward that extreme. The example of the average number of sit-ups by 10 subjects discussed earlier in this chapter illustrates this phenomenon.

The precision of the mean should not exceed the precision of the data by more than one decimal place. For example, if distance is measured in meters, it is appropriate to calculate the mean to the nearest tenth of a meter, or decimeter, but not to the nearest thousandth of a meter, or millimeter. Convention permits the precision of the mean to be one significant figure beyond the accuracy of the data.

Calculating the Mean

The mean is computed simply by summing all the scores and dividing the sum by the number of scores (N). In statistics, summing scores is represented by the symbol Σ, which is the uppercase figure for the Greek character sigma. This symbol is read as "the sum of." For example, if a variable is labeled X, then ΣX should be read "the sum of X."

To denote the mean of a variable, we place a line over the symbol for the variable. In equation 4.01, X is the symbol for the variable, so the mean of X is represented by $\overline{X}$, often read as "X-bar." If Y represents the variable, $\overline{Y}$, or Y-bar represents the mean of Y.

The formula for the mean of the variable X is

$$\overline{X} = \frac{\Sigma X}{N}.$$ (4.01)

From a Rank Order Distribution

Computing the mean or average of a small set of scores is easily done: Add all the scores and divide the sum by the number of scores.

X
33.9
31.2
29.6
27.0
24.7
25.0
23.0

$$\Sigma X = 194.4$$

$$\overline{X} = \Sigma X/N = 194.4/7 = 27.77$$

The value the calculator returns for 194.4/7 is 27.77142857. But the data are significant only to the nearest tenth, so the mean value is calculated to the nearest hundredth. Rounding the mean to one significant figure beyond the data provides users of the data with the information that the mean is closer to 27.8 than to 27.7.

From a Simple Frequency Distribution

The same principle is applied for calculating the mean from a simple frequency distribution, but the formula is modified so that each score is multiplied by its frequency, to account for all of the subjects who received a given score. The formula thus becomes

$$\overline{X} = \frac{\Sigma(fX)}{N}. \tag{4.02}$$

This equation requires that each raw score (X) be multiplied by its frequency (f). The values of all fX are then summed to obtain the total of all scores. The number of scores (N) is represented by the sum of the frequency column. Table 4.1 presents an application of equation 4.02 for a set of data.

Table 4.1   Calculation of the Mean From a Simple Frequency Distribution

X	f	fX
10	1	10
9	4	36
7	6	42
5	7	35
3	4	12
2	3	6
1	3	3
	$N = 28$	$\Sigma(fX) = 144$

$$\bar{X} = 144/28 = 5.1$$

From a Grouped Frequency Distribution

The equation for calculating the mean from a grouped frequency distribution is the same as for a simple frequency distribution, but X is replaced by the midpoint of each group:

$$\bar{X} = \frac{\Sigma(fX)_{mid}}{N}, \tag{4.03}$$

where X_{mid} is the midpoint of each group.

Although it is essential that a researcher know the techniques for calculating the mean by hand, the process is often tedious and time-consuming. Computers perform these functions much more rapidly and with greater accuracy. Computers can also remember large amounts of information, so it is not necessary for a computer to group the data. When one knows how to do the calculations by hand on simple numbers for small values of N—and understands the meaning of the answers—then the computer can add speed, accuracy, and convenience.

Relationships Among the Mode, Median, and Mean

When the data are distributed normally, the three measures of central tendency all fall at or near the same value. But when the data are skewed, these measures are no longer identical. Figure 4.1 demonstrates the relationships among the three measures on a positively skewed distribution.

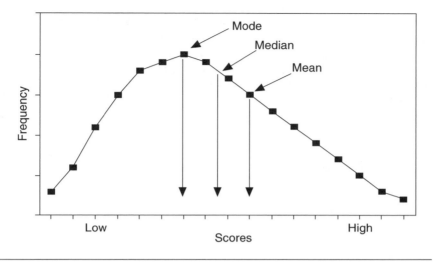

Figure 4.1   Relationships of measures of central tendency on a positively skewed curve.

Note that the highest scores are farther from the mode than are the lowest scores. This characteristic of positively skewed curves shifts the median and the mean to the right of the mode. On a positively skewed curve the three measures of central tendency read from left to right in the following order: mode, median, mean. On a negatively skewed curve the order is reversed: mean, median, mode.

When deciding which of the three measures to use, consider the following:

- Use the mode if only a rough estimate of the central tendency is needed and the data are normal.
- Use the median if (a) the data are on an ordinal scale, (b) the middle score of the group is needed, (c) the most typical score is needed, or (d) the curve is badly skewed by extreme scores.
- Use the mean if (a) the curve is near normal and the data are of the interval or ratio type, (b) all available information from the data is to be considered (i.e., the order of the scores as well as their relative values), or (c) further calculations, such as standard deviations or standard scores, are to be made.

Summary

Measures of central tendency are values that describe the central characteristics of a set of data. It is essential to know the central tendency of a set of scores in order to evaluate any of the raw scores in the set. The three measures of central tendency are the mode, the median, and the mean. The mode provides a quick estimate of central tendency for all types of data. The median can be used on ordinal data or when the data are badly skewed. The mean is the preferred measure of central tendency for interval or ratio data that are normally or almost normally distributed. When the data are normally distributed, the three measures of central tendency all fall at or near the same value.

Problems to Solve

1. Calculate the mean for each of the following sets of data:

A.	X		B.	X	f
	17			17	2
	15			16	4
	13			15	5
	12			14	7
	12			13	7
	11			12	8
	11			11	10
	9			10	9
	8			9	5
	8			8	3
	8				
	5				
	4				
	2				
	2				

See appendix C for answers to problems.

Key Words in This Chapter

Central tendency
Mode
Median

Terminal statistic
Mean
Outlier

Chapter 5

Measures of Variability

Consider the theoretical scores of two golfers, one a beginner and one a professional. Each hits 100 balls toward the pin with a 9-iron. The pro is very consistent, and all of the pro's balls lie close to and equally spread out around the pin. The beginner is inconsistent; some shots land close to the pin, others far to either side, still others far beyond or far short of the pin. If we measured the distance of every ball from the pin (consider shots that fell short or left to be negative values and those that fell long or right to be positive), then the average, or mean, of each set of values could be near zero, because negative left balances positive right, and negative short balances positive long. If only the average distance of the balls from the pin for both players were evaluated, we might erroneously conclude that they were equal in ability, because each average distance from the pin would be near zero. But when the variability of their performances is reviewed, the pro is clearly a much more consistent golfer than the beginner.

 $\mathbf{V}$ ariability is a measure of the spread, or dispersion, of a set of data. Once the mean, median, and mode have been calculated, the central tendencies of the raw data are known. However, data can be either compact around the central measure or spread out. *Leptokurtic curves* typically represent limited variability, whereas *platykurtic curves* represent large variability. Consequently, it is important to know the variability, or spread, of the group as well as its central tendency.

When both central tendency and variability are known, two or more sets of data can be completely compared. It is often interesting to compare the means of two or more sets of data to determine which set has a higher average, but it may be more important to compare the variability of the data because the variabilities may differ while the means can be similar.

Most statisticians recognize four main measures of variability: range, interquartile range, variance, and standard deviation. Of these four, the variance and the standard deviation are the most important because they lead to additional calculations that may answer further questions about the data. The range and interquartile range are limited in their applicability, but they may serve as useful rough estimates of variability that are easy to calculate.

Range

Recall from chapter 2 that the range is simply the difference between the highest and lowest scores in the data set. This is the best quick estimate we can make of the variability of data. It is easily calculated, but it is only a very rough measure. The range is somewhat analogous to the mode in that it is only an estimate of variability and is very unstable; it can change radically if extreme scores are introduced to the data set.

Interquartile Range

Another measure of variability often used when data are ordinal, or when ratio or interval data are highly skewed, is called the **interquartile range** or *IQR*. This range is the difference between the raw scores at the 75th (Q_3) and the 25th (Q_1) percentile points. It is calculated as follows:

$$IQR = Q_3 - Q_1 . \tag{5.01}$$

The *IQR* is a useful measure of the spread of data if the investigator is more interested in the middle scores than in the extremes. Because the interquartile range considers only 50% of the data (the scores that fall in the middle half of the data set), it is not affected by highly divergent scores at the extremes. This is why it is useful for skewed data. Like the median, it presents a typical picture, but it does not consider all information available about the data.

An alternate measure of variability when using percentiles is the **semi-interquartile range** (*SIQR*), which is half of the *IQR*. The formula is $SIQR = (Q_3 - Q_1)/2$. This measure is used if a smaller indicator of the variability is desired.

Variance

Both the range and the *IQR* consider only two points of data in determining variability. They do not include the values of the scores between the high and low, or between the Q_1 and Q_3 data points. Another way to assess variability, which does consider the values of each data point, is to determine the distance of each raw score from the mean of the data. This distance is called deviation and is denoted by *d*.

The computation of deviation scores for two sets of data, each of which has a mean of 25, is demonstrated in table 5.1.

The sum of all deviations around the mean must be zero in both the *X* and the *Y* examples. Regardless of the size of the scores, the sum of the deviations around any mean is zero. This is one way to verify the accuracy of the mean. If the deviations do not sum to zero, the mean is incorrect.

In the second example (*Y*), the deviation scores are larger and the range is larger ($R_X = 27 - 23 = 4$ and $R_Y = 35 - 15 = 20$). In fact, as a comparison of the ranges shows, the *Y* data are five times more variable than the *X* data. The interquartile range comparisons confirm this conclusion: ($IQR_X = 26 - 24 = 2$ and $IQR_Y = 30 - 20 = 10$). This relationship between the range and the *IQR* will always be consistent when the data from the two sets are both normally distributed.

If we sum the absolute values of the deviations (ignoring the direction of the deviation) from the mean, we find that the total deviation of *X* is 6, and the total deviation of *Y* is 30, or five times more variable than *X*.

Table 5.1   Deviations From the Mean

X	d	d_{ABS}	Y	d	d_{ABS}
27	+2	2	35	+10	10
26	+1	1	30	+5	5
25	0	0	25	0	0
24	−1	1	20	−5	5
23	−2	2	15	−10	10
$\Sigma X = 125$	$\Sigma d = 0$	$\Sigma d_{ABS} = 6$	$\Sigma Y = 125$	$\Sigma d = 0$	$\Sigma d_{ABS} = 30$

$$\bar{X} = 125/5 = 25$$
$$\bar{Y} = 125/5 = 25$$

In these examples, however, the signs of the deviations have meaning; they indicate whether the raw score is above or below the mean. Because we need to know this, we cannot ignore the signs without losing information about the data. There is an algebraically acceptable way to eliminate the negative signs: If we simply square the deviations, the squared values are all positive.

The average of the squared deviations provides another method of determining the variability of the data set. This method is more useful than the range or interquartile range because it considers each score and its distance from the mean in determining variability. The **variance** is the average of the squared deviations from the mean. The symbol V is used for variance in this text; other texts may use the symbol S^2. Variance is represented in algebraic terms in equation 5.02:

$$V = \frac{\Sigma d^2}{N}.$$
(5.02)

Table 5.2 shows how the variance is determined for the previous examples of X and Y.

Standard Deviation

The calculation of variance shown in table 5.2 suggests that Y is 25 times more variable than X ($50/2 = 25$), whereas we previously concluded from the range and interquartile range that Y was only 5 times more variable than X. This discrepancy is the result of squaring the deviation scores. To bring the value for the variance into line with other measures of variability (and with the unit values of the original raw data), we compute the square root of the variance.

Table 5.2   Calculation of Variance

X	d	d^2	Y	d	d^2
27	+2	4	35	+10	100
26	+1	1	30	+5	25
25	0	0	25	0	0
24	−1	1	20	−5	25
23	−2	4	15	−10	100
$\Sigma X = 125$	$\Sigma d = 0$	$\Sigma d^2 = 10$	$\Sigma Y = 125$	$\Sigma d = 0$	$\Sigma d^2 = 250$

$$V_X = 10/5 = 2$$
$$V_Y = 250/5 = 50$$

The resulting value is called the standard deviation because it is standardized with the unit values of the original raw data. The **standard deviation** is the square root of the average of the squared deviations from the mean. This definition applies to a population of scores (the standard deviation of a sample is discussed later in this chapter).

The Definition Method

The standard deviation of a population is symbolized by the Greek lowercase sigma (σ) and is represented algebraically as follows:

$$\sigma = \sqrt{\frac{\Sigma d^2}{N}}. \tag{5.03}$$

In the example from table 5.2, the standard deviation of X is

$$\sigma_x = \sqrt{\frac{10}{5}} = \sqrt{2} = 1.414,$$

and the standard deviation of Y is

$$\sigma_Y = \sqrt{\frac{250}{5}} = \sqrt{50} = 7.071.$$

These values are now consistent with the range and IQR because the standard deviation of Y is five times as large as the standard deviation of X ($1.414 \times 5 = 7.07$). This statistic gives an accurate and mathematically correct description of the variability of the group while considering each data point and its deviation from the mean. The standard deviation is very useful because many advanced statistical techniques are based on a comparison of the mean as the measure of central tendency and the standard deviation as the measure of variability.

The method just described for calculating the standard deviation is called the definition method; it is derived from the verbal definition of standard deviation. But this is a cumbersome and lengthy procedure, especially when N is large and the mean is not a whole number. Under these conditions there is a high probability of mathematical error during the calculation.

When the data are in a frequency distribution, the deviation scores (d) must be multiplied by their respective frequencies (f). The formula for application with a simple frequency distribution is

$$\sigma = \sqrt{\frac{\Sigma f d^2}{N}}. \tag{5.04}$$

Applying this formula to a large problem results in some lengthy calculations. Table 5.3 demonstrates how equation 5.04 is applied to the data from chapter 3, table 3.2, on basketball free throw attempts.

Note that the mean must be calculated before the values for the deviation column can be determined. Because this unwieldy method can easily result in calculation errors, it is almost never used in practical statistics. It is presented here to demonstrate the theoretical calculation of standard deviation. Other equations, which are derived from the definition equation, give the same answer but are easier to use. They are presented in the next section.

The Raw Score Method

The difficulty encountered in the definition method results from the many decimal figures in the deviation scores when the mean is not a whole number. With algebraic manipulation, the formula can be restated in terms that use raw scores only. This raw score formula makes the calculations easier and requires fewer columns.

The definition formula (5.03),

$$\sigma = \sqrt{\frac{\Sigma d^2}{N}}$$

Table 5.3   Calculation of Standard Deviation by the Definition Method From Simple Freqency Data

X	f	fX	d	d^2	fd^2
10	2	20	3.583	12.838	25.676
9	6	54	2.583	6.672	40.031
8	9	72	1.583	2.506	22.553
7	12	84	0.583	0.340	4.079
6	15	90	−0.417	0.174	2.608
5	8	40	−1.417	2.008	16.063
4	4	16	−2.417	5.842	23.368
3	2	6	−3.417	11.676	23.352
2	1	2	−4.417	19.510	19.510
1	1	1	−5.417	29.344	29.344
	N = 60	ΣfX = 385			Σfd² = 206.584

$$\overline{X} = \frac{\Sigma fX}{N} = \frac{385}{60} = 6.417$$

$$\sigma = \sqrt{\frac{\Sigma fd^2}{N}} = \sqrt{\frac{206.584}{60}} = 1.86$$

Note. The mean and *d* values are calculated to three decimal places to minimize errors when squaring rounded numbers.

can be modified by expansion because $d = (X - \overline{X})$. Then

$$\Sigma d^2 = \Sigma(X - \overline{X})^2$$

and

$$\Sigma(X - \overline{X})^2 = \Sigma X^2 - 2\Sigma X\overline{X} + \Sigma\overline{X}^2.$$

Also, because summing scores is the same as multiplying their mean times N,

$$\Sigma X = N\overline{X} \text{ and } \Sigma\overline{X}^2 = N\overline{X}^2.$$

Therefore,

$$\Sigma d^2 = \Sigma X^2 - 2N\overline{X}\overline{X} + N\overline{X}^2 \text{ or } \Sigma d^2 = \Sigma X^2 - 2N\overline{X}^2 + N\overline{X}^2.$$

Combining like terms, we obtain

$$\Sigma d^2 = \Sigma X^2 - N\overline{X}^2.$$

Then, we substitute in the original formula (equation 5.3) so that

$$\sigma = \sqrt{\frac{\Sigma X^2 - N\overline{X}^2}{N}} \text{ or } \sigma = \sqrt{\frac{\Sigma X^2}{N} - \frac{N\overline{X}^2}{N}}$$

and cancel values of N to obtain

$$\sigma = \sqrt{\frac{\Sigma X^2}{N} - \overline{X}^2}. \tag{5.05}$$

Equation 5.05 is much simpler to apply because it requires only the squaring and summing of the raw scores, which is why it is sometimes referred to as the raw score formula.

The raw score formula may be applied to any of the three data sets we have previously described: rank order distributions, simple frequency distributions, and grouped frequency distributions. However, because grouped frequency distributions almost always represent large Ns, which should be entered into a computer, only the formulas to calculate standard deviation from rank order and simple frequency distributions will be demonstrated.

Rank Order Distributions

An exercise physiologist collected resting heart rate data on cross-country runners. To calculate the standard deviation of the data, the researcher applied equation 5.05 as shown in table 5.4.

Table 5.4   Calculation of Standard Deviation by the Raw Score Method From Rank Order Data

X	X^2
65	4,225
60	3,600
58	3,364
42	1,764
51	2,601
61	3,721
$\Sigma X = 337$	$\Sigma X^2 = 19,275$

$$\overline{X} = \frac{337}{6} = 56.2$$

$$\sigma = \sqrt{\frac{19,275}{6} - (56.2)^2} = 7.4$$

Simple Frequency Distributions

If simple frequency distribution data are used, the formula is modified to include f:

$$\sigma = \sqrt{\frac{\Sigma(fX^2)}{N} - \overline{X}^2} \tag{5.06}$$

Table 5.5 shows how equation 5.06 can be used to calculate the standard deviation of the basketball free throw data from table 5.3. The calculations in table 5.5 use the raw scores only and do not use the deviations.

Calculating Standard Deviation for a Sample

Recall from chapter 1 that most research is performed on samples taken from populations. The sample is always smaller than the population and is selected randomly so that the statistics calculated on the sample are representative of the corresponding parameters in the population. The researcher assumes that conclusions based on the sample are applicable to the population from which the sample was drawn.

Samples rarely contain the extreme values that are found in the population. For example, if we randomly sampled the weights of 100 men in a university with 15,000 male students, it is not likely that anyone in the sample would weigh 350 pounds or 100 pounds, although there may be students in the population who weigh these amounts. The variability of the sample is almost never as large as the variability of the population.

Table 5.5   Calculation of Standard Deviation by the Raw Score Method From Simple Freqency Data

X	f	fX	X^2	fX^2
10	2	20	100	200
9	6	54	81	486
8	9	72	64	576
7	12	84	49	588
6	15	90	36	540
5	8	40	25	200
4	4	16	16	64
3	2	6	9	18
2	1	2	4	4
1	1	1	1	1
	$N = 60$	$\Sigma fX = 385$		$\Sigma fX^2 = 2{,}679$

$$\overline{X} = \frac{385}{60} = 6.417 *$$

$$\sigma = \sqrt{\frac{2{,}679}{60} - (6.417)^2} = 1.86$$

* Calculated to three significant decimals to minimize errors when squaring.

When standard deviation is calculated from a sample and then used to estimate the standard deviation of the population, a correction factor must be applied to the equation so that the estimate of the population is not biased by a small sample. Without this correction factor, an estimate of a population standard deviation based on a sample would be erroneously small.

The equations presented previously in this chapter for calculating standard deviation are based on the assumption that an entire population has been measured. If these equations were applied to samples, an error when generalizing from a sample to a population would occur. The correction to the equations is based on the concept of degrees of freedom.

The **degrees of freedom** are the number of values in a data set that are free to vary. If no restrictions are placed on the data, then all values are free to vary; that is, they may take on any value, limited only by the precision of the measuring instrument and the actual values in the population. But when we take a sample and make the assumption that the sample represents the population, a restriction is placed on the data in the sample.

When the sample mean is assumed to be identical to the population mean (i.e., the sample mean is established), the sum of the sample data is also set because the mean equals the sum divided by N. The sum must be a value that yields a mean equal to the population. This limits the numerical value of the last data point in the sample to a number that will create a sample statistic theoretically equal to the population parameter (even though the population value is unknown).

Assume that the mean of four values must be 5.0. Therefore, the sum of the four numbers must equal 20. Let the values 2, 3, and 7 be the first three numbers. What must the fourth number equal to bring the sum to 20? It must be 8. The last number is not free to vary; it is limited to only one value, that which will create a sum of 20. Therefore, this example has 3 degrees of freedom. Three of the four numbers can assume any value, but the last one must be whatever it takes to make the sum equal 20. The degrees of freedom for a single data set that is a sample representing a population are always $N - 1$. In this example, df $= N - 1 = 4 - 1 = 3$.

This correction factor, when applied to the definition formula for standard deviation (equation 5.3) reduces the denominator by 1. The standard deviation for a sample is

$$SD = \sqrt{\frac{\Sigma d^2}{N-1}}. \tag{5.07}$$

Note that the symbol for standard deviation in equation 5.07 is SD rather than σ. In this book, we use SD to represent the standard deviation for a sample and σ to represent the standard deviation for a population. (In some statistics books, standard deviation may be represented by S or s.)

Like equation 5.03, equation 5.07 is difficult to apply when N is large and the mean is not an integer, and it is almost never used to calculate standard deviations. But it may be modified so that deviation scores are not required. The modified formula is

$$SD = \sqrt{\frac{\Sigma X^2 - \frac{(\Sigma X)^2}{N}}{N-1}}. \tag{5.08}$$

Equation 5.08 is commonly used to calculate standard deviation from samples arranged in a rank order distribution. If we apply equation 5.08 to the data from table 5.4, the result is

$$SD = \sqrt{\frac{19,275 - \frac{(337)^2}{6}}{6-1}} = 8.3.$$

The standard deviation calculated by equation 5.08 yields a slightly larger answer ($SD = 8.3$) than does equation 5.05 ($\sigma = 7.4$). The difference represents the correction for degrees of freedom. If the data in table 5.4 are assumed to represent a sample from a larger population, the estimate of the population standard deviation would be 8.3.

For use on simple frequency data, equation 5.08 may be modified by adding f. The resulting equation is

$$SD = \sqrt{\frac{\Sigma fX^2 - \dfrac{(\Sigma fX)^2}{N}}{N-1}}. \tag{5.09}$$

When we apply equation 5.09 to the data in table 5.5, the calculation becomes

$$SD = \sqrt{\frac{2,679 - \dfrac{385^2}{60}}{60-1}} = 1.88.$$

The standard deviation obtained using equation 5.09 ($SD = 1.88$) is only slightly larger than the one calculated using equation 5.06 ($\sigma = 1.86$) because N is relatively large (60). As N increases, the differences between the standard deviations obtained with the sample and the population formulas decrease. When $N \geq 50$, the differences are usually insignificant.

Choosing the Appropriate Formula

Several equations have been presented for calculating standard deviation. All result in similar answers when applied to the same data. The choice of which equation to use depends on (a) the organization of the data and (b) whether the data represent a sample or a population.

The definition equations (equations 5.03, 5.04, and 5.07) are almost never used. They are presented in this chapter only to explain the theoretical base for standard deviation. If these equations are used at all, they are used only on rank order data when the scores are small and the mean is a whole number.

If the data represent a population, the raw score equations (equations 5.05 and 5.06) may be used. They are especially easy to use when the mean is already known. This is usually the case, because an investigator who is interested in the standard deviation is also interested in the mean.

When N is large (≥ 50), a computer is almost always used to calculate the standard deviation. Most computer programs will produce both the population value (σ) and the sample value (SD). If only one value is produced, it is usually SD.

When the data represent a sample, the appropriate equation for rank order data is equation 5.08, and for simple frequency data, equation 5.09. You should be familiar with the application of each of these equations.

Summary

Standard deviation is the most common method of measuring the variability, or dispersion, of a data set. It considers the deviation of each data point and its distance from the mean. When we compare several different sets of data using the standard deviations of each, we can determine the relative spread of the sets. When the standard deviation is small, the group is compact. When it is large, there is more diversity, or spread, among the scores in the group.

When N is large and the distribution is close to normal, there are usually 5 or 6 standard deviations within the range of a data set. That is, there are about 3 standard deviations from the mean to the largest score and from the mean to the smallest score (see figure 5.1).

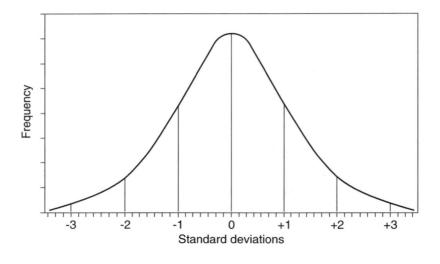

Figure 5.1 Distribution of standard deviations on a normal data set with large N.

If N is small or the data are skewed, this rule does not hold. When N is large and the range contains either fewer than or more than about 5 or 6 standard deviations, the researcher would be wise to check the calculations for error.

Standard deviation and variance are commonly used concepts in statistics. You should be familiar with both their calculation and their meaning. In later chapters, important techniques for determining relationships among variables (correlation) and for comparing differences between group means (t tests and analysis of variance) are discussed. Variance and standard deviation are critical factors in these techniques.

Problems to Solve

In an exercise physiology laboratory, a student collected data on the sum of skinfold thicknesses in millimeters at the triceps and subscapular sites. The following measurements were obtained: 17, 19, 12, 24, 26, 18, 15, 14, 20.

1. What is the standard deviation if the data represent a population?
2. What is the standard deviation if the data represent a sample?
3. What is the variance for the population?
4. Assume the following set of data on basketball shots completed in 1 minute represents a sample from a larger population. Calculate the standard deviation using an appropriate formula.

X	f
29	3
27	2
26	4
25	10
24	13
23	7
22	7
21	2
20	1

See appendix C for answers to the problems.

Key Words in This Chapter

Variability
Interquartile range
Semi-interquartile range

Variance
Standard deviation
Degrees of freedom

Chapter 6

The Normal Curve and Sampling Error

How good is a 14-foot long jump? Without additional information, we can't be sure. If the jump is completed by a male high school student during track practice, we can compare it to the jumps of teammates. If we know that the average of all long jumpers on the team is 17'1/2", then we know that this score is below the mean, but how much? To evaluate a raw score, we must compare it to a scale that has known central tendency and variability. Scores from such scales are called standard scores. Standard scores provide information that helps us evaluate the raw score.

$\mathbf{W}$hen studying the concept of variability, especially while learning about standard deviation, students often ask, What does standard deviation mean? They know how to calculate it, but they do not understand its meaning or its value. The answer lies in an understanding of the unique characteristics of the **normal curve.**

Observed over a sufficient number of cases, many variables will assume a normal distribution. This is fortunate for the statistician because the characteristics of the normal curve are well known. If the data are from an interval or a ratio scale, if enough cases have been measured, and if the curve is normal or nearly normal, the characteristics of normality may be applied to the data.

Figure 6.1 shows a frequency polygon of data on the weights of a large group ($N > 1,000$) of college-age males. The mean weight is 175 pounds, and the standard deviation is 25 pounds. The graph shows that most subjects weigh between 150 and 200 pounds ($\bar{X} \pm 1\sigma$) and that the highest frequency is at the mean (175 pounds). A few subjects weigh less than 100 pounds, and a few weigh more than 250, but not many.

Most subjects' weights are near the mean. As values progress farther from the mean in either direction, fewer and fewer cases are represented. The standard deviation units are distributed equally above and below the mean, and the majority of the cases fall between the mean and $\pm 1\sigma$, or between 150 and 200 pounds.

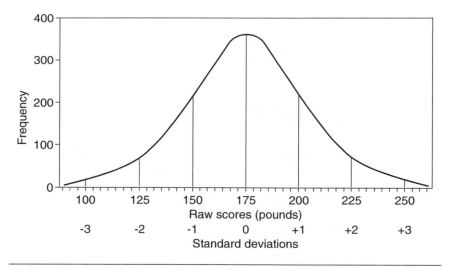

Figure 6.1 Relationship between raw scores and Z scores on a normal curve.

Z Scores

A **Z score** is a raw score expressed in standard deviation units. If the standard deviation of the scores in figure 6.1 is 25, then 1 standard deviation unit is equivalent to 25 pounds on the raw score scale. A score of 200 lies 25 raw score units, or 1 standard deviation unit, above the mean. The raw score of 200 is equivalent to a Z score of +1. Likewise, a raw score of 150 (25 raw units, or 1 standard deviation unit, below the mean) has a Z score of –1.

When the mean and the standard deviation of any set of normal data are known, Z can be calculated for any raw score (X) by using the following formula:

$$Z = \frac{X - \overline{X}}{\sigma}. \tag{6.01}$$

NOTE: If the data represent a sample, use S.D. in the denominator.

Using this formula, we can calculate that a male student who weighs 200 pounds has a Z score of +1:

$$Z = \frac{200 - 175}{25} = +1.$$

When the raw score is less than the mean, the Z score is negative. If the raw score is 165,

$$Z = \frac{165 - 175}{25} = -0.4.$$

A unique characteristic of the normal curve is that the percentage of area under the curve between the mean and any Z score is known and constant. When the Z score has been calculated, the percentage of the area (which is the same as the percentage of raw scores) between the mean and the Z score can be determined.

In any normal curve, 34.13% of the scores lie between the mean and one Z score in either the positive or negative direction. Therefore, when we say that most of the population falls between the mean and ±1 Z score, we are really saying that 68.26% (2 × 34.13 = 68.26), or about two-thirds, of the population falls between these two limits. This is true for any variable on any data provided that the distribution is normal. Figure 6.2 demonstrates this concept.

Table A.1 in appendix A may be used to determine the percentage of scores that falls between the mean and any given Z score. The numbers on the left and right sides of the table represent Z scores to the nearest tenth of a point. The values across the top represent Z scores to the nearest hundredth.

To determine the percentage of scores that falls between the mean and ±1.00 Z score we proceed down the left-hand column to the value 1.0 and move across the row to the .00 column; the value in that column is 34.13. This value is the percentage of raw scores that falls between the mean and either +1.00 Z or –1.00 Z. For ±2.00 Z the value in table A.1 is 47.72 (which is equal to 34.13 + 13.59, the percentages of the area from 0 to +1 and from +1 to +2 in figure 6.2). This indicates that 47.72% of the population of scores lies between the mean and +2.00 Z scores, or between the mean and –2.00 Z scores.

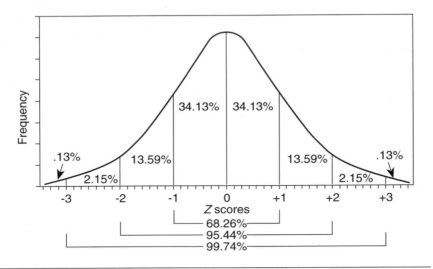

Figure 6.2 Percentage of area under the normal curve for selected Z score values.

Doubling that number (47.72 × 2 = 95.44) tells us that slightly more than 95% of all raw scores lies between the mean and ±2 Z. The figure for 3 standard deviations is 99.74% (49.87 × 2 = 99.74). This confirms that most raw scores fall within ±3 standard deviations of the mean (see figure 6.2).

Because the normal curve is bilaterally symmetrical, table A.1 in appendix A provides only the values for one-half of the curve; the values are the same for positive and negative Z scores. The percentage corresponding to a Z score of –1.78 is found by proceeding down the left-hand column to 1.7, then across to the .08 column, where the value 46.25 is read. This is interpreted to mean that 46.25% of the population of scores lies between the mean and 1.78 Z scores in either direction.

Table A.1 may also be used for the opposite procedure, to find the Z score that corresponds to a given percentage of the population. If we want to know the Z value that represents 30% of the area under the curve, we look for the figure of 30.00 in the body of the table. This exact number cannot be found, so we find the closest value to 30.00 (29.95). This value corresponds to a Z score of ±0.84. So to be 30 percentage points above or below the mean, a person must have a raw score equal to about ±0.84 Z scores.

Converting Z Scores to Percentile Scores

Once we know the Z score for a raw score, we can also determine the percentile value of that score by looking in table A.1 in appendix A. Because the mean, or a Z score of 0.00, represents the 50th percentile, any figure read from the table for a positive Z score is added to 50 to determine the percentile of that score. A Z score of +1.24 has a corresponding table value of 39.25. Therefore the raw score equivalent to a Z score of +1.24 has a percentile value of 89.25 (50 + 39.25 = 89.25). If the Z score is negative (–1.24), then the value from the table is subtracted from 50. This results in a percentile score of 10.75 (50 – 39.25 = 10.75).

Just as any raw score has a corresponding Z score, so each Z score has a corresponding percentile score. Table A.1 may be used to convert scores from Z to percentiles, or vice versa. It's important to remember that a positive Z score must be added to 50, and a negative one must be subtracted from 50, to determine the percentile value of the raw score. The values in table A.1 represent only half of the curve; they should be interpreted as the distance from the middle of the curve toward either end.

Standard Scores

Raw scores are the direct result of measuring a value, usually distance, time, force, or frequency. Raw scores may have different units of measurement, different mean values, and different variability. A **standard score** is derived from raw data and has a known central tendency and variability. Only when raw scores are converted to the same standard score base can they be compared directly.

It is not logical to compare 50 sit-ups in 1 minute with a mile-run time of 4:36.3. Which performance is better, 20 feet in the long jump or 10.5 seconds in the 100-meter dash? We cannot answer such questions using raw data alone, because the values are based on different units of measurement. To make an appropriate comparison, we must first convert raw scores to one of the four standard scores: percentiles, Z scores, T scores, or stanines.

Percentiles

In chapter 3, we discussed the percentile. This type of standard score has several advantages. Percentiles have 50% as their central tendency and 0% to 100% as their range. They also have known quartile and decile divisions.

We can easily compare different types of raw scores when we convert them to percentiles. If a 5'5" high jump represents the 75th percentile, and a time of 11.5 seconds in the 100-meter dash represents the 80th percentile, then the running score is better than the jumping score. These scores are now directly comparable because they have both been converted to the same base, or standard.

Z Scores

The calculation of Z scores was explained earlier in this chapter; now we will discuss how to use Z scores as standard scores. Z scores have a known central tendency ($\bar{Z} = 0$) and known variability ($\sigma = 1.0$). When raw scores are converted to Z scores, two or more sets of data may be directly compared. For example, which score is better, a Z score on a long jump test of -1.3 or a Z score of -0.50 on a gymnastics floor exercise test? The gymnastics score is better, because a Z score of -0.50 is higher (thus better) than -1.3.

We can confirm this by using table A.1 in appendix A to convert both scores to percentiles: -1.3 Z equals a percentile score of 9.68 ($50 - 40.32 = 9.68$), and -0.5 Z equals a percentile score of 30.85 ($50 - 19.15 = 30.85$). The student who received these scores is not a whiz at either the long jump or floor exercise, but it is safe to say that the student is a better performer in gymnastics. Figure 6.3 presents a graphic representation of the two scores.

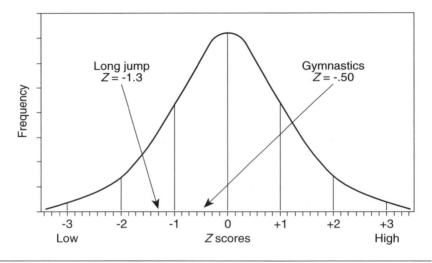

Figure 6.3   Comparison of Z score values on the normal curve.

T Scores

A third standard score, called the T score, is often used to report norms in educational settings, such as those on national fitness or skill tests. By convention and definition, a **T score** has a mean of 50 and a standard deviation of 10. A T score of 60 is 1 standard deviation above the mean, and a T score of 30 is 2 standard deviations below the mean. Before T can be calculated, the corresponding Z score must be known. Then the formula for converting from Z to T scores is

$$T = 10\,Z + 50.$$ (6.02)

The T equivalent for a Z score of +1.5 is 65.0 ($T = 10 \times 1.5 + 50 = 65.0$). When the Z score is negative, the T score will be less than 50 (remember that Z scores have a mean of 0). In the previous example of gymnastics and long jump scores, the T score for a Z score of -0.50 in gymnastics is calculated as follows: $T = 10 \times (-0.50) + 50 = 45.0$.

The T scale was created because the lay public has difficulty understanding Z scores. It is not common to think of scores with 0 as the mean. Most people consider 0 to be nothing and prefer to have 50 as the middle and a range from 0 to 100. This is why percentiles are so widely used; they are easy to understand.

T scores have a mean of 50 and a range from 0 to 100, but it is very unlikely that a T score would be less than 20 or greater than 80 because these figures represent Z scores of -3 and $+3$, respectively. As figure 6.2 shows, only 2×0.13, or 0.26%, of the population lies beyond the ± 3 limit on the Z scale. Indeed, a T score of 100 would represent a Z score of +5 and a percentile of 99.99999, which is a rather unlikely occurrence.

Figure 6.4 shows the relationships among Z scores, percentiles, and T scores.

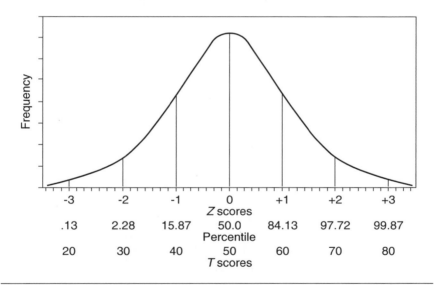

Figure 6.4   Relationships among Z scores, percentiles, and T scores on the normal curve.

Stanines

Like the T scale, the **stanine** scale (a derivation of the words standard nine) is commonly used for reporting the results of educational tests. Parents who inquire about their children's scores on standardized tests may find the results presented in stanines. For example, a student may score at the 7th stanine in math, the 4th stanine in reading, and the 5th stanine in verbal skills. Physical education teachers need to understand stanines so they can read the reports found in the student files.

How are stanine scores interpreted? As figure 6.5 shows, in the stanine scale, the standard normal curve is divided into nine sections with 5 as the middle score, 1 as the lowest score, and 9 as the highest score.

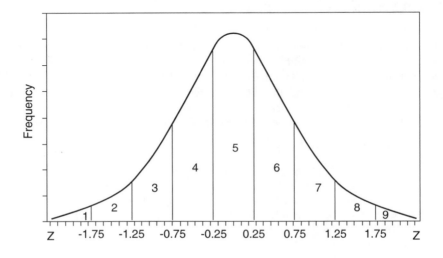

Figure 6.5   Stanine distribution on the normal curve.

To calculate a stanine score, we need to know the raw score, the mean, and the standard deviation. From this information, we calculate the Z score. Once we have the Z score, we can find the stanine score directly from figure 6.5.

Each section on the stanine curve is one-half Z score wide except for stanines 1 and 9. The center section (stanine 5) ranges from -0.25 Z to $+0.25$ Z. The other sections continue to the left or right in 0.5 Z intervals.

Scores that fall exactly on a dividing line between 2 stanines are usually given the higher value. Thus, a Z score of $+0.75$ is considered to be in the 7th stanine, and a Z score of -1.25 is in the 3rd stanine.

Stanine scores do not represent an exact raw score. Rather they represent a range, or section of the curve, into which the raw or Z score falls. In this way, stanines are similar to quartiles or deciles. When only the stanine is known, it is impossible to tell exactly where a raw score falls within the stanine. Only the section of the curve that best represents the score is known.

A student who scored at the 7th stanine on pull-ups, the 4th stanine on the mile run, and the 5th stanine on the sit-and-reach test of flexibility is considerably above the mean in strength, slightly below the mean in aerobic capacity, and very close to the mean in flexibility.

Predicting Population Parameters Using Statistical Inference

In many sciences, little research is conducted on entire populations. Often the population is so large that it would be impossible to measure each member. In such cases, the researcher takes a sample of the population and assumes that the sample represents the population and that the characteristics of the sample statistics are indicative of the population parameters. For this assumption to be valid, the sample must be randomly selected. For more information on sampling selection, see chapter 1.

Earlier we discussed the problem of determining the mean weight of all men at a university. If there were 100 or fewer men in the population, we could measure all of them, and a sample would not be needed. But if the population of all men in the university were 15,000, it would be too time-consuming to measure all of them. So we would take a random sample to *estimate* the mean of the population.

The sample size is limited by such factors as time restraints, lack of finances, and facilities and equipment. If we wanted to measure height or weight, a large sample could be collected because it is easy to measure these variables. But if we were interested in hydrostatically measured body composition or $\dot{V}O_2$max on a treadmill or bicycle ergometer, perhaps only a few subjects could be measured.

Let us assume that a sample size of 50 men is desired. If the administration permitted us access to the student files, a computer could select 50 males from the files in a completely random fashion. These randomly selected subjects could then be invited to participate in the study and could be measured by appointment. (It would be wise to invite more than 50 subjects because a few may not participate.) The mean weight of this sample could then be used to represent the mean weight of all men at the university.

Estimating Sampling Error

Sampling error refers to the amount of error in the estimate of a population parameter that is based on a sample. Even though a sample is randomly drawn, it is unlikely that the mean of the sample will be identical to the mean of the population. Also, the true population mean is never known exactly because all members of the population are never measured. Consequently, we need a way to determine how accurate the sample mean is and what the odds are that it is deviant from the population mean by a given amount.

The **standard error of the mean** is a numeric value that indicates the amount of error that may occur when a random sample mean is used as a predictor of the mean of the population from which the sample was drawn. By accepting this error, we admit that the exact population mean can never be known, unless every member of the population is measured. We can only know the sample mean and how much error it is estimated to have.

The prediction of the population mean is always an educated guess and is accompanied by a probability statement. That is, the population mean is assumed to exist between some set limits, and the chance of this assumption being correct is stated as odds such as 90 to 10, 95 to 5, or 99 to 1.

Let us use the example of estimating mean weight of men at a university to explain the theory behind this technique. We assume that the range of weight of all men in the population is about 100 to 250 pounds and that the mean is about 175 pounds (a value we can never know precisely unless we measure all 15,000 men).

Suppose a large number (theoretically an infinite number) of random samples is taken with 50 subjects in each sample. After each sample is taken, the subjects are returned to the population pool so that they have an equal chance of being chosen again in a subsequent sample. Most samples have means between 165 and 185 pounds and ranges of about 125 to 225 pounds. The range of the population (which we estimated to be 100 to 250 pounds) will be larger than the range of any one sample because it is unlikely that the extremes of a population of 15,000 will be randomly selected into a sample of 50.

If a true random sample of sufficient size is taken each time, the sample mean will not vary greatly from the actual population mean. It is unlikely that a random sample of $N = 50$ would have a mean value near one of the extremes of the population, because for this to happen, all subjects in the sample would have to be from one extreme of the population. But this unusual occurrence is more likely to happen if the sample size is small (e.g., 5). To avoid this potential error, we make our samples as large as possible within the limits of our resources. *The probability of obtaining a biased sample increases as the sample size decreases. The larger the sample, the smaller the error in predicting the population mean.*

After all of the samples have been taken and the mean of each calculated, the sample means could be arranged into a graph that would approximate a normal curve. The means of this series of random samples would be normally distributed, even if the population from which they were drawn is not normal. This concept is known as the **central limit theorem.**

Each sample has its own mean and standard deviation, and the total group of sample means also has a mean (the mean of the means) and a standard deviation (the standard deviation of the means). This value, the standard deviation of the means, is called the **standard error of the mean.**

Because the sample means form a normal distribution, all of the characteristics of normality apply to the curve of the sample means. The mean of the sample means becomes the best estimate of the true population mean because it represents the results of a large number of randomly drawn samples of 50 subjects each.

Let us assume that the mean of the sample means is 175 pounds, and the standard deviation of the sample means, or the standard error of the mean, is 3.5. Applying our knowledge about the relationship of percentile points to Z scores on a normal curve (see figures 6.1 and 6.2), and assuming that the mean of the sample means is the best estimate of the population mean, we can state that the population mean probably lies somewhere between 175 ±1 × 3.5 pounds (171.5 to 178.5).

Recall that approximately 68% of the area under any normal curve lies between ±1 standard deviation (see figure 6.2), and that the remaining 32% lies outside these limits (16% on each end). The mean of the means (175) becomes the estimate of the population mean, and there is a slightly better than 2-to-1 chance (actually it is 68 to 32) that the estimate of 175 ±1 × 3.5 pounds is correct.

If we widen the population estimate (making it less precise but more encompassing) to ±2 standard deviations (175 ±2 × 3.5) and estimate that the population mean lies somewhere between 168 and 182, we increase the odds of being correct to better than 95 to 5 (95.44% of the area under the normal curve lies between ±2 standard deviations of the mean; see figure 6.2). And if we make the estimate accurate to ±3 standard deviations (175 ±3 × 3.5) and estimate that the population mean is between 164.5 and 185.5 pounds, then the odds that our estimate is correct are better than 99 to 1.

Using this technique, we can make probability statements about the population mean with various degrees of accuracy. *The more precise, or narrow, the estimate, the lower the odds of being correct. As the estimate becomes more general, or broad, the odds of being correct improve.*

The process just described is intended to show the theory behind the concept of the standard error of the mean. In practice, it may be as difficult to measure a large number of samples of 50 subjects each as it would be to measure all 15,000 males in the population. Fortunately, an equation has been derived that will estimate the standard deviation of a series of theoretical sample means based on one random sample only.

The standard error of the mean estimated from one sample is denoted by SE_M. This value is algebraically demonstrated by equation 6.03:

$$SE_M = \frac{SD}{\sqrt{N}}, \tag{6.03}$$

where SD is the standard deviation of the sample and N is the sample size. (Refer to equation 5.08 for the formula for SD.)

Using only one randomly drawn sample and equation 6.03, we can calculate the odds that the population mean lies within certain limits. In the example we have been discussing, suppose we take one random sample of 50 subjects with a mean of 175 pounds and standard deviation of 25 pounds. Then we can calculate that SE_M is equal to 3.5:

$$SE_M = \frac{25}{\sqrt{50}} = 3.5.$$

This value is actually a standard deviation on a normal curve; therefore, it is equivalent to a Z score of ±1.0. We may infer from this calculation that the mean of the population from which this sample was drawn has a 68% chance of being within the limits of 175 ±1 (3.5) pounds. The process of inference is represented by the following equation:

$$\mu = \overline{X} \pm 1\,(SE_M),$$

where $\overline{X}$ represents the sample mean and μ (mu in the Greek alphabet) represents the population mean. In our example, this equation would indicate μ is 175 $\pm$ 1(3.5), or somewhere between 171.5 and 178.5. This is sometimes written as 171.5 $\leq \mu \leq$ 178.5. A similar formula for calculating the standard error of a proportion is available when data are presented as percentiles (see discussion on the t test for proportions in chapter 8).

Levels of Confidence and Probability of Error

A **level of confidence** (LOC) is a percentage figure that establishes the probability that a statement is correct. It is based on the characteristics of the normal curve. In the previous example, the estimate of the population mean (μ) is accurate at the 68% level of confidence because we included one SE_M (i.e., 1Z) above and one SE_M below the predicted population mean.

But if there is a 68% chance of being correct, there is also a 32% chance of being incorrect. This is referred to as the **probability of error** and is written $p <$ .32 (the probability of error is less than .32). The area under the normal curve that represents the probability of error is called **alpha** (α). Alpha is the level of chance occurrence. In statistics, this is sometimes called the error factor (i.e., the probability of being wrong because of chance occurrences that are not controlled). Alpha is directly related to Z because it is the area under the normal curve that extends beyond a given Z value.

Remember that SE_M is a standard deviation on a normal curve. By including 2 standard errors above and below μ (175 $\pm$2 $\times$ 3.5, or 168 to 182 pounds), we increase our level of confidence from 68% to better than 95% and drop the error factor from 32% to about 5%.

To be completely accurate when we use the 95% level of confidence, or p = .05, we should not go quite as far as 2 Z scores away from the mean. In table A.1 in appendix A, the value in the center of the table that represents the 95% confidence interval is 47.50 (95/2 = 47.50, because table A.1 represents only half of the curve). This corresponds to a Z score of 1.96. The value 1.96 is the number of Z scores above and below the sample mean that accurately represents the 95% level of confidence, or $p = .05$. The correct estimate of μ at $p = .05$ is 175 ±1.96 × 3.5, or 175 ± 6.9 pounds (168.1 to 181.9).

A similar calculation could be made for the 99% level of confidence by looking in table A.1 to find the value that reads 49.5 (99/2 = 49.5). This exact value is not found in the table. Because 49.5 is halfway between 49.49 and 49.51 in the table, we choose the higher value (49.51), which gives us slightly better odds. The Z score correlate of 49.51 is 2.58. To achieve the 99% level of confidence, we multiply SE_M by ±2.58. The estimate of the population mean at the 99% level of confidence ($p = .01$) is 175 ±2.58 × 3.5 or 175 ± 9.0 pounds. This may be expressed as $166 \le \mu \le 184$, $p = .01$.

Likewise, we could establish the 90% level of confidence by looking up 45% (90/2 = 45) in table A.1. The Z score correlate of 45% is 1.65, so μ is 175 ±1.65 × 3.5, or 175 ± 5.8 pounds, $p = .10$.

The level of confidence (chances of being correct) and probability of error (chances of being incorrect) always add to 100%, but by tradition, the level of confidence is reported as a percent and the probability of error (p) is reported as a decimal. The Z values to determine p at the most common levels of confidence are listed in table 6.1. Other values may be determined for any level of confidence by referring to table A.1 in appendix A.

The generalized equation for determining the limits of a population mean based on one sample for any level of confidence is presented as equation 6.04.

$$\mu = \overline{X} \pm Z(SE_M), \tag{6.04}$$

where Z is a Z score that will produce the desired probability of error (i.e., $Z = 1.65$ for $p = .10$, 1.96 for $p = .05$, and 2.58 for $p = .01$).

Table 6.1   Corresponding Values for Z, LOC, and p

Z	LOC	p
1.0	68%	.32
1.65	90%	.10
1.96	95%	.05
2.58	99%	.01

LOC = level of confidence.
p = probability of error (sum of both tails of the curve).

An Example Using Statistical Inference

A researcher was interested in the average height of first-grade children in a school district. Eighty-three students randomly selected from throughout the district were measured with the following results: $\bar{X} = 125$ centimeters and $SD = 10$ centimeters. The population height was estimated at the 95% level of confidence, $p = .05$, as follows:

$$SE_M = \frac{10}{\sqrt{83}} = 1.1$$

and

$$\mu = 125 \pm 1.96\,(1.1) = 125 \pm 2.2, \quad p = .05$$

The researcher concluded with 95% confidence that the mean height of all the first-grade children in the school district was between 122.8 and 127.2 centimeters ($122.8 \le \mu \le 127.2$). There is, however, a 5% chance ($p = .05$) that this conclusion is incorrect.

In kinesiology, as in most behavioral sciences, the most common minimum level of confidence used is 95% ($p = .05$). But there is a developing trend to accept some research at the 90% ($p = .10$) level. The researcher decides which level to use, but the reader of the research must be the ultimate judge of what is acceptable. By consulting table A.1 in appendix A, we can determine any level of confidence and its equivalent Z. The decision of which level to use is based on the consequences of being wrong.

In medical research, if an incorrect conclusion may result in serious injury or death to the patient, then a very high level of confidence is desired. Even the 99% ($p = .01$) level may not be sufficient. The reader may wonder, Why not always use $p = .01$, since it is the least likely to result in error? Because, while the prediction of the population mean is more accurate at $p = .01$ than $p = .05$, it is more broad, thus less precise. In statistics, if you want less error, you must sacrifice precision. When an incorrect conclusion will not result in bodily harm or excessive financial loss, lower levels may be used. The user of the research must determine the consequences of being wrong in each case and accept or reject the conclusions accordingly. Franks and Huck (1986) provide an excellent review of the history and procedures for selecting a level of confidence. Chapter 8 discusses the pros and cons of selecting levels of confidence that are too low or too high.

Calculating Skewness and Kurtosis

A major assumption of the previous discussion about statistical inference is that the characteristics of the normal curve can be applied. Consequently, it is critical that we know if the data deviate from normality. **Skewness** is a measure of the bilateral symmetry of the data, and **kurtosis** is a measure of the relative peakedness of the curve of the data.

By observing the graph of the data and identifying the three measures of central tendency, we can get a general idea of the skewness of the data, but this method is not exact (see figure 4.1). Using Z scores, we can obtain a numerical value that indicates the amount of skewness or kurtosis in any set of data.

Because Z scores are a standardized measure of the deviation of each raw score from the mean, we can use Z scores to determine if the raw scores are equally distributed around the mean. When the data are completely normal, or bilaterally symmetrical, the sum of the Z scores above the mean is equal but opposite in sign to the sum of the Z scores below the mean. The positive and negative values cancel each other out, and the grand sum of the Z scores is zero.

If we take the third moment (the cube of the Z scores, or Z^3), we can accentuate the extreme values of Z, but the signs of the Z values remain the same. This places greater weight on the extreme scores and permits a numeric evaluation of the amount of skewness. Computing the average of the Z^3 scores produces a raw score value for skewness. The formula for calculating the raw value for skewness is

$$\text{Skewness} = \frac{\Sigma Z^3}{N}. \tag{6.05}$$

When the Z^3 mean is zero, the data are normal. When the Z^3 mean is positive, the data are skewed positive, and when the Z^3 mean is negative, the data are skewed negative. This effect can be seen by examining the data presented in table 6.2. Notice that the data are skewed negative. When these data are graphed (see figure 6.6), the skewness is easily observed.

Table 6.2 Calculation of Skewness and Kurtosis

X	Z score	Z^3	Z^4
5	1.08	1.26	1.36
5	1.08	1.26	1.36
5	1.08	1.26	1.36
4	0.27	0.02	0.01
4	0.27	0.02	0.01
4	0.27	0.02	0.01
4	0.27	0.02	0.01
4	0.27	0.02	0.01
3	−0.54	−0.16	0.09
3	−0.54	−0.16	0.09
2	−1.36	−2.52	3.42
1	−2.17	−10.22	22.17
		$\Sigma Z^3 = -9.18$	$\Sigma Z^4 = 29.90$

Mean = 3.67
SD = 1.23
N = 12

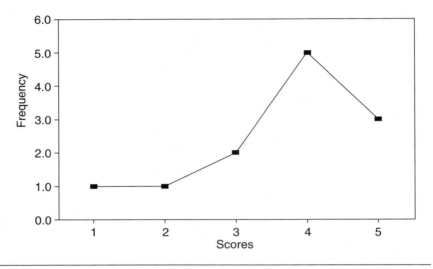

Figure 6.6   Negative skew.

Kurtosis may also be calculated from Z scores. By taking the fourth moment (Z^4) of the Z scores, the extreme Z values are again accentuated, but the signs are all converted to positive. When the average of the Z^4 value is 3.0, the curve is normal. To make the units equal for both skewness and kurtosis, the mean of Z^4 is typically reduced by 3.0. The formula for calculating the raw value for kurtosis is (AndersonBell, 1989, p. 173; Spiegel, 1961, p. 91)

$$\text{Kurtosis} = \left(\frac{\Sigma Z^4}{N} \right) - 3.0. \tag{6.06}$$

A score of 0 indicates complete normal kurtosis, or a mesokurtic curve, just as a score of 0 for skewness indicates complete bilateral symmetry. When the raw score for kurtosis is greater than 0.0, the curve is leptokurtic (more peaked than normal), and when the raw score is less than 0.0, the curve is platykurtic (more flat than normal).

Raw skewness and kurtosis scores are not easily interpreted, because a raw score alone does not indicate a position on a known scale. But when raw scores are converted to Z scores, they are easy to interpret. To convert the raw scores for skewness (equation 6.05) or kurtosis (equation 6.06) to Z scores for skewness or kurtosis, we divide the raw scores by the standard error. According to Dixon (1990, p. 137) the standard error for skewness is

$$SE_{skew} = \sqrt{\frac{6}{N}},$$

and the standard error for kurtosis is

$$SE_{kurt} = \sqrt{\frac{24}{N}}.$$

If we divide the raw scores for skewness or kurtosis by the appropriate standard error, we obtain a Z_{skew} or Z_{kurt} value:

$$Z_{skew} = \frac{\Sigma Z^3 / N}{\sqrt{6/N}},$$ (6.07)

$$Z_{kurt} = \frac{\Sigma Z^4 / N - 3.0}{\sqrt{24/N}}.$$ (6.08)

These values may be interpreted as Z scores (i.e., values greater than 1.96 or less than -1.96 exceed $p = .05$, and values greater than 2.58 or less than -2.58 exceed $p = .01$). Typically, data are considered to be within acceptable limits of skewness or kurtosis if the Z values do not exceed ± 2.0.

Using the data from table 6.2, we can find Z_{skew} in the following manner:

$$\text{Skewness} = \frac{-9.18}{12} = -0.77$$

$$SE_{skew} = \sqrt{\frac{6}{12}} = 0.71$$

$$Z_{skew} = \frac{-0.77}{0.71} = -1.08$$

Z_{kurt} can be found as follows:

$$\text{Kurtosis} = \frac{29.90}{12} - 3.0 = -0.51$$

$$SE_{kurt} = \sqrt{\frac{24}{12}} = 1.41$$

$$Z_{kurt} = \frac{-0.51}{1.41} = -0.36$$

From these values ($Z_{skew} = -1.08$ and $Z_{kurt} = -0.36$) we can determine that the data in table 6.2 and figure 6.6 are slightly skewed negative and slightly platykurtic;

however, neither value approaches significance (±2.0). So we may conclude that the data are within acceptable ranges of normality. Note that data sets with small values of N may appear to be significantly skewed when graphed (see figure 6.6), but the true evaluation of the degree of skewness must by made by Z score analysis.

Summary

Raw scores may be converted to standard scores in the form of Z, percent, T, or stanine to provide more information about the data and to assist in evaluating raw data. Standard scores are also useful for comparing the results of tests measured on different units of measurement (e.g., comparing time to force or distance). Because standard scores have a common central tendency and variability, data presented in standard score units may be directly compared regardless of the unit of measurement of the raw score.

By using the characteristics of the normal curve, estimates of the parameters of populations may be made from sample statistics. Calculations can also be performed to determine the standard error of the mean, which indicates the amount of error in an estimate of a population mean based on a random sample.

The assumption underlying the process of statistical inference is that the characteristics of the normal curve can be applied to the data. We can determine how much the distribution of a data set deviates from normality by calculating both skewness and kurtosis.

Problems to Solve

1. Find the percentage of values that falls between the mean of a given set of data and a Z score of +0.35.
2. Given a set of data with a mean of 25.7 and a standard deviation of 5.2, calculate the Z score equivalents of the following raw scores: (a) 21.6, (b) 28.9, and (c) 24.5. What is the interpretation of the Z scores?
3. Use table A.1 in appendix A to determine the equivalent percentile rank of each of the scores in problem 2.
4. Convert the Z scores calculated in problem 2 to T scores.
5. What is the stanine value for each of the scores in problem 2?

6. Using the following sample data, compute the standard error of the mean. What can you say about the value of SE_M in relation to the size of the SD and N?

	SD	N
a.	21.4	36
b.	2.1	106
c.	56.7	19

7. Suppose you wanted to know the mean $\dot{V}O_2$max in milliliters per kilogram per minute of all females at a university. You randomly selected 50 names from the files of the administration office and invited the subjects to be tested in the lab. Thirty-five accepted the invitation and were tested. The mean $\dot{V}O_2$max of this sample was calculated to be 38.5 with a standard deviation of 4.7. If you wanted the population mean estimate to be accurate at the 95% level of confidence, what would be the upper and lower limits of the predicted $\dot{V}O_2$max value?

8. Using the best random sample you can get, collect data and estimate the average height of all members of your class. Select the appropriate level of confidence. If time permits, measure the total population (all members of the class). How close did your estimate come to the true population mean?

See appendix C for answers to problems.

Key Words in This Chapter

Normal curve	Central limit theorem
Z score	Level of confidence
Standard score	Probability of error
T score	Alpha
Stanine	Skewness
Sampling error	Kurtosis
Standard error of the mean	

Chapter 7

Correlation, Bivariate Regression, and Multiple Regression

In an exercise physiology class, a student learns that the time it takes to run 1.5 miles is directly related to oxygen consumption as measured in the laboratory by $\dot{V}O_2max$ (ml $\times$ kg^{-1} $\times$ min^{-1}). The instructor explains that subjects with the highest $\dot{V}O_2max$ values will tend to have the fastest times in the 1.5-mile run. In statistical terms, the instructor is saying these two variables are correlated. Knowing that two scores are correlated allows prediction (using a technique called linear regression) of the value of one variable using the other variable. In this case, knowing how fast someone can complete a 1.5-mile run will allow an estimate of the person's $\dot{V}O_2max$. How is this prediction made?

Correlation

Correlation is widely used in kinesiology. For example, kinesiologists might calculate the correlation between height and weight or the correlation between skinfold thicknesses and percent body fat. **Correlation** is defined as a numerical coefficient that indicates the extent to which two variables are related or associated—that is, the extent to which the direction and size of deviations from the mean in one variable are related to the direction and size of deviations from the mean in another variable. This technique is referred to as **Pearson's product moment correlation coefficient;** it is named after Karl Pearson (1857–1936), who developed the current version in 1896 in England (Kotz & Johnson, 1982, p. 199).

The **coefficient,** or number, that represents the correlation is always between +1.00 and −1.00. A perfect positive correlation (+1.00) would exist if every subject varied an equal distance from the mean in the same direction (measured by a Z score) on two different variables. For example, if every subject who was 1 Z score above the mean on variable X was also 1 Z score above the mean on variable Y, and every other subject showed a similar relationship between deviation score on X and deviation score on Y, the resulting correlation would be +1.00.

Similarly, if all subjects who were above or below the mean on variable X were an equal distance in the opposite direction from the mean on variable Y, the resulting correlation would be −1.00. If these interrelations are similar but not perfect, then the correlation coefficient is less than +1.00, such as .90 or .80 in the positive case, and greater than −1.00, such as −.90 or −.80 in the negative case. A correlation coefficient of 0.00 means that there is no relationship between variables.

Positive correlations result when subjects who receive high numerical scores on one variable also receive high numerical scores on another variable. For example, there is a positive correlation between the number of free throws taken in practice and the percentage of free throws made in games over the season. Players who spend lots of practice time on free throws tend to make a high percentage of free throws in games.

Negative correlations result when scores on one variable tend to be high numbers and scores on a second variable tend to be low numbers. The relationship between distance scores and time scores is almost always negative because greater distance (e.g., a far long jump) is associated with faster runners (e.g., low time scores).

Correlation is a useful tool in many types of research. In evaluating testing procedures, correlation is often used to determine the validity of the measurement instruments by comparing a test of unknown validity with a test of known validity. It may also be used as a tool for prediction. When the correlation coefficient between two variables is known, scores on the second variable can be predicted based on scores from the first variable.

Although a correlation coefficient indicates the amount of relationship between two variables, it does not indicate the cause of that relationship. Just because one variable is related to another, that does not mean that changes in one will cause changes in the other. Other variables may be acting on one or both of the related variables and affecting them in the same direction.

For example, there is a positive relationship between IQ score and collegiate grade point average (GPA), but a high IQ does not always result in a high GPA. Other variables—motivation, study habits, financial support, study time, parental and peer pressure, and instructor skill—affect grades. The IQ score may account for some of the variability in grades (explained variance), but it does not account for all of the variability (unexplained variance). Therefore, although there is a relationship between IQ and grades, it cannot be said that grades are a direct function of IQ. Other unmeasured variables may influence the relationship between IQ and grades, and this unexplained variance is not represented by the single correlation coefficient between IQ and grades.

Cause and effect may be present, but correlation does not prove cause. Changes in one variable will not automatically result in changes in a correlated variable, although this may happen in some cases. The length of a person's pants and the length of his or her legs are positively correlated. People with longer legs have longer pants. But increasing one's pant length will not lengthen one's legs!

A visual description of a correlation coefficient may be presented as a **scatter plot,** in which the scores for each subject on two variables are plotted with one variable on the X-axis and the other variable on the Y-axis. A scatter plot in figure 7.1 shows the graph of the relationship between the long jump and the triple jump (hop, step, and jump). Because both require running speed, and the triple jump contains a long jump, we would expect that the two variables are related—that is, people who are good long jumpers will probably be good triple jumpers also, and vice versa.

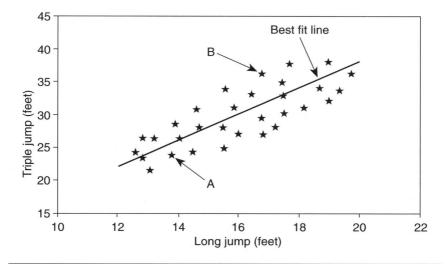

Figure 7.1   Positive correlation.

Each star represents the plot of a person's score on both variables. Subject A long-jumped about 14 feet and triple-jumped about 24 feet. Subject B had scores of about 17 feet and 36 feet, respectively. The other stars represent the scores of additional subjects.

Note how all of the stars seem to cluster around the line. This line is known as the **best fit line;** it represents the best linear estimate of the relationship between the two variables. In figure 7.1, the correlation between the two variables is about +.80.

The best fit line for a negative correlation slopes in the opposite direction. Figure 7.2 shows a scatter plot for the relationship between time in the 100-meter dash and distance in the long jump measured in meters. Low scores in the 100-meter dash (good performance) tend to be associated with high scores in the long jump (good performance), and vice versa. This makes sense, because we know that an important factor in long jump ability is the speed at takeoff. Generally speaking, the faster one runs, the farther one jumps.

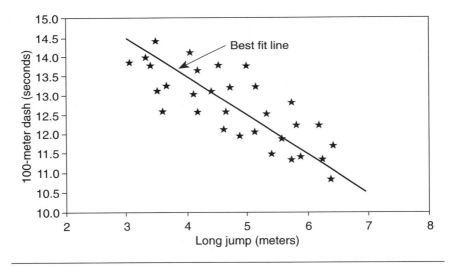

Figure 7.2   Negative correlation.

The value of the correlation coefficient in figure 7.2 is about $-.80$. The data plots form around the best fit line, which decreases in Y value as it increases in X value. This is characteristic of a negative correlation.

The positive or negative sign of the coefficient does not indicate the strength or usefulness of the correlation coefficient. It is the absolute numerical size of the coefficient, not its direction, that determines the strength or usefulness of the relationship. As the correlation gets closer to +1.00 or −1.00, the plots cluster closer to the line, and the relationship becomes stronger. If the correlation were a perfect +1.00 or −1.00, all of the plotted points would fall exactly on the line. As the correlation approaches 0.00, the plot drifts away from the line until there is no best fit line. A 0.00 correlation plot has points all over the graph and they form the shape of the entire graph; the points do not approach the form of a line.

With a 0.00 correlation, it is not possible to predict one variable from another. A 0.00 relationship, or one close to it, is called orthogonal, meaning that the plot is square or rectangular rather than linear. An orthogonal relationship is one that is not statistically significant.

Figure 7.3 shows a scatter plot of the correlation between long jump distance and grade point average. This graph has no best fit line. The plot assumes a rectangular rather than a linear shape. We cannot predict a person's grade point average from his or her distance on the long jump. The plots are more dense between GPAs of 2.0 and 3.0 because most grades are between C and B.

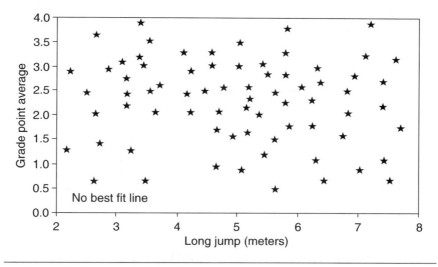

Figure 7.3   Zero correlation.

Be aware that sometimes the best fit line is not straight. Two variables may be related in a **curvilinear** fashion. Pearson's correlation coefficient is a measure of linear relationships. If it is applied to curvilinear data, it may produce a spuriously low value with a corresponding interpretation that the relationship is weak or non-existent when in fact the relationship is strong but not linear.

Figure 7.4 shows a scatter plot for data that have a curvilinear relationship. The curved line represents the true relationship between the X and Y variables. The straight line represents the relationship assumed by Pearson's coefficient. The true relationship is curvilinear and strong. The spurious linear relationship is weak and incorrect. It is important to examine the scatter plot of the data in addition to calculating the correlation coefficient. The pattern of scores on the scatter plot can help us decide whether a linear measure is appropriate, or whether a nonlinear model might better explain the relationship.

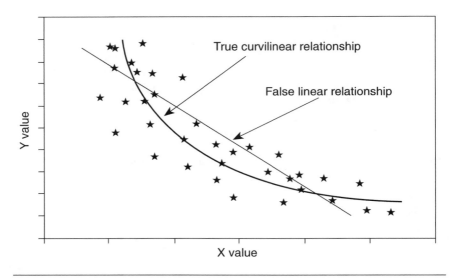

Figure 7.4   Comparison of curvilinear and linear relationships.

Calculation of the relationship between nonlinear variables is possible but will not be discussed here. Determining the shape of the relationship is called trend analysis. This procedure is discussed in more advanced textbooks; a good discussion of trend analysis is found in Keppel (1982, chapter 7).

Calculating the Correlation Coefficient

There are several methods of determining the value of Pearson's correlation coefficient (r). We first consider the definition formula because it is derived from the definition of correlation stated at the beginning of this chapter. Recall that a correlation coefficient represents the relationship between the Z scores of the subjects on two different variables (usually designated X and Y). This can be stated mathematically as the mean of the Z score products for all subjects.

The definition formula for Pearson's product moment correlation coefficient (r) is

$$r = \frac{\Sigma(Z_X Z_Y)}{N},$$

(7.01)

where Z_x and Z_y are the Z scores for each subject on the X and Y variables, and $N =$ the number of pairs of scores.

To determine r with this formula, we first compute the mean and standard deviation of each variable and then compute the Z score for each subject on each variable. We will demonstrate this method with simple data from table 7.1.

After determining the means and standard deviations, we calculate Z_X and Z_Y for each subject. Then we multiply Z_X by Z_Y for each subject and sum the products. The correlation coefficient r then becomes

$$r = \frac{3.787}{4} = .947.$$

Table 7.1 uses equation 5.05 (the equation for populations) to calculate standard deviation. The population formula is used in table 7.1 rather than the sample formula because the sample formula is intentionally biased by the placement of $N - 1$ in the denominator to correct for the use of a sample as a predictor of a population. Correlation is not an inferential statistic. It simply states the relationship between the two variables for a given N. Therefore it would be inappropriate to bias the standard deviation by using $N - 1$ in the denominator. The population formula for standard deviation is necessary for an accurate calculation of r.

The mean and standard deviation values in table 7.1 are carried to three significant places more than the data (which are reported only to the nearest integer value) would seem to permit. This is done to insure accuracy in the resultant r

Table 7.1   Calculation of r by the Definition Formula

X	Y	X^2	Y^2	Z_X	Z_Y	$Z_X Z_Y$
2	1	4	1	−1.347	−1.342	1.808
3	2	9	4	−.577	−.447	.258
5	3	25	9	.962	.447	.430
5	4	25	16	.962	1.342	1.291
$\Sigma X = 15$	$\Sigma Y = 10$	$\Sigma X^2 = 63$	$\Sigma Y^2 = 30$			$\Sigma Z_X Z_Y = 3.787$

$$\bar{X} = \frac{15}{4} = 3.75$$

$$\bar{Y} = \frac{10}{4} = 2.5$$

$$\sigma X = \sqrt{\frac{63}{4} - 3.75^2} = 1.299$$

$$\sigma Y = \sqrt{\frac{30}{4} - 2.5^2} = 1.118$$

$$r = \frac{+3.787}{4} = +.947$$

value. When means are multiplied together, or when standard deviations are squared, the error of rounding is also squared.

The value obtained for r from the data in table 7.1 is positive and close to 1.0. By looking at the data, we can confirm that the values of the X and Y pairs are related. The lowest X (2) is paired with the lowest Y (1), and the highest values of X (two tied at 5) are paired with the highest Y (4) and the second highest Y (3).

Also, the Z score products in table 7.1 are all positive. This is because the values for X that are below the mean of X (and have negative Z scores) are paired with values for Y that are below the mean of Y (and have negative Z scores). The product of two negative numbers is positive. Likewise, the values of X above the mean are paired with values of Y that are above the mean. The products of these pairs are also positive. Therefore, the sum of $Z_X Z_Y$ is positive. If the Z score pairs are not of the same sign, the Z score products will be negative. When low X values are associated with high Y values, and vice versa, the negative Z score products sum to a negative value, producing a negative value for r.

If the Z score products are mixed, some positive and some negative (indicating no special relationship among the X and Y pairs), the positive and negative products tend to cancel each other out, and the sum of $Z_X Z_Y$ is small, approaching zero. This produces a very low value for r.

Understanding the concept of relationships between X and Y pairs is critical to the comprehension and use of correlation. You should study the example in table 7.1 until these relationships become clear.

Because it is a laborious process to use, equation 7.1 is not applied in most problems. Applying this formula to large values of N, especially when the raw values and means are not whole numbers, takes considerable time and effort. Equation 7.1 is presented here only to provide a theoretical understanding. Two other formulas, which have been derived from the definition formula but are much easier to apply, are presented next.

The Mean/Standard Deviation Formula

An alternate formula for determining r, the mean/standard deviation formula, requires only the means and standard deviations of each variable and the cross products of the raw scores. This formula may appear more complicated than the definition formula, but it is easier to apply with raw data because the Z scores do not have to be calculated. Note that the standard deviation values are based on equation 5.05 for a population, not equation 5.08, which is for samples only. The mean/standard deviation formula is

$$r = \frac{\dfrac{\Sigma XY}{N} - \overline{X}\,\overline{Y}}{\sigma_X \sigma_Y}. \tag{7.02}$$

Table 7.2 demonstrates the use of this formula to determine the relationship between height in inches (X) and weight in pounds (Y) of 10 college-age men and women. Using the data in table 7.2, we calculate the means,

$$\overline{X} = \frac{676}{10} = 67.6 \quad \text{and} \quad \overline{Y} = \frac{1,471}{10} = 147.1,$$

and the standard deviations,

$$\sigma_X = \sqrt{\frac{45,940}{10} - (67.6)^2} = 4.92 \quad \text{and}$$

$$\sigma_Y = \sqrt{\frac{239,649}{10} - (147.1)^2} = 48.23.$$

With these values, we can determine r:

$$r = \frac{\dfrac{101,555}{10} - (67.6)\,(147.1)}{(4.92)\,(48.23)} = .892.$$

The resulting value for r of .892 represents a substantial relationship between height and weight. The high positive correlation can be confirmed by looking at the data in table 7.2. The shortest person is also the lightest (61 inches and

Table 7.2   Calculation of r by the Mean/Standard Deviation Formula

X	Y	X^2	Y^2	XY
72	167	5,184	27,889	12,024
64	114	4,096	12,996	7,296
71	135	5,041	18,225	9,585
66	140	4,356	19,600	9,240
63	115	3,969	13,225	7,245
62	106	3,844	11,236	6,572
73	207	5,329	42,849	15,111
61	102	3,721	10,404	6,222
76	260	5,776	67,600	19,760
68	125	4,624	15,625	8,500
$\Sigma X = 676$	$\Sigma Y = 1,471$	$\Sigma X^2 = 45,940$	$\Sigma Y^2 = 239,649$	$\Sigma XY = 101,555$

X = height in inches.
Y = weight in pounds.

102 pounds), and the tallest is the heaviest (76 inches and 260 pounds). Other scores tend to line up in the same order on height as they do on weight.

Because the data include both men and women, the r value is probably higher than would be found if only men or only women were measured. The combined group has a greater range of both height and weight than does either a group of men or a group of women. Because average height and weight are greater for men than for women, the men tend to cluster at the top of each scale and the women at the bottom, thus raising the r value.

The Machine Formula

An alternate formula for determining r is sometimes preferred because it does not require calculating means and standard deviations. Raw sums of the X, Y, X^2, Y^2, and XY columns are all that is needed. Because it uses nonrounded values, this method is slightly more accurate. (In the mean/standard deviation formula, rounded means and standard deviations distort the answer slightly.)

This formula is sometimes referred to as the machine formula or raw score formula because it uses only sums of columns and is often the formula that is built into calculators and computers. It is calculated as follows:

$$r = \frac{N(\Sigma XY) - (\Sigma X)(\Sigma Y)}{\sqrt{[N(\Sigma X^2) - (\Sigma X)^2][N(\Sigma Y^2) - (\Sigma Y)^2]}}. \tag{7.03}$$

When the machine formula is applied to the data from table 7.2, it yields essentially the same results as were obtained with the mean/standard deviation formula (equation 7.02):

$$r = \frac{10(101,555) - (676)(1,471)}{\sqrt{[10(45,940) - (676)^2][10(239,649) - (1,471)^2]}} = .891.$$

The difference of .001 in the answer is due to the squaring of the error produced by rounding when the 2 standard deviations are multiplied together in the denominator of equation 7.02. The answer from equation 7.03 (.891) is correct, whereas the answer from equation 7.02 is in error by .001.

Selecting the Correct Formula

The definition formula (equation 7.01) is almost never used in practical research. When used with large values of N and noninteger means, this formula presents a high probability of arithmetic or recording errors. The choice of which of the other two formulas to use depends on the availability of means and standard deviations.

The machine formula should be used if the means and standard deviations are not available or are not needed for other calculations. If the means and standard deviations are accurately known from prior calculations, then the choice of formula is a matter of personal preference.

Evaluating the Size of the Correlation Coefficient

As discussed earlier, the correlation is perfect if r is +1.00 or −1.00. But how useful is a correlation of .9 or −.85 or .3? One of the main purposes of correlation is the prediction of one variable from the value of another variable for a given subject. For predictive purposes, absolute r values lower than about .7 may produce unacceptably large errors in an individual prediction, especially if the standard deviations of either or both variables are large. As a general rule, an absolute r value of .5 to .7 is considered low, .7 to .8 is moderate, and .9 or higher is good for predicting Y from X. Values lower than .5, if they are statistically significant, can be useful for identifying nonchance relationships between variables, but they are probably not large enough to be useful in predicting individual scores.

It is sometimes helpful to represent the relationship between two variables with a diagram in which each variable is depicted by a circle. If the circles do not overlap, there is no relationship. If they overlap completely (one directly on top of the other), the correlation is +1.00.

If the circles overlap somewhat, as in figure 7.5, the area of overlap represents the amount of variance in the dependent, or predicted, variable that can be explained by the independent, or predictor, variable.

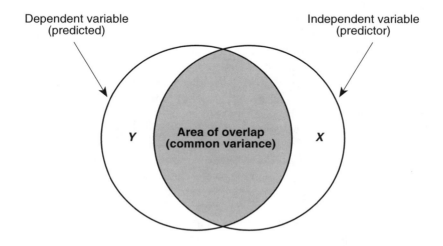

Figure 7.5   Venn diagram for a bivariate relationship.

The area of overlap, called the common variance, is equal to r^2. If two variables are correlated at $r = .8$, then they have 64% common variance ($.8^2 = .64$). This means that 64% of the variability in the Y variable can be explained by variance in the X variable. The remaining 36% of the variance in Y is unexplained. This unexplained variance is responsible for error when predicting Y from X. For example, strength and speed are related at around $-.80$. This means that they have 64% common variance. That is, 64% of both strength and speed come from common factors. The remaining 36% is unexplained; that is, it is not accounted for by the correlation coefficient.

Determining the Significance of the Correlation Coefficient

The usefulness of the coefficient is based on the size and significance of r. If r is reliably different from 0.00, the r value is significant; it did not result from a chance occurrence. If r is significant, we can conclude, with known odds of being correct, that the relationship between the variables is real—that is, it is caused by a factor or factors other than chance. This implies that if we measured the same variables on another set of similar subjects, we would get the same r value (i.e., the correlation is reliable).

When N is small (for example, 3 or 4 pairs of scores), it is possible that a spuriously high r value can occur by chance. Suppose the numbers 1, 2, and 3 are written on small pieces of paper and placed in a hat. The numbers are blindly drawn one at a time on two different occasions. It is possible that the numbers could be drawn in the same order twice. This would produce an r value of $+1.00$. But this r value would not be real, because no factors other than chance would be operating on the variables to cause the relationship.

In contrast, the odds of 100 numbers being randomly selected in the same order twice are very low. Therefore, if the r value with $N_{pairs} = 100$ is high, we conclude that chance is not a factor because when N is large, it is rare for a high r to occur by chance. Some factor other than chance must cause the relationship. A large N makes us more confident that a high r value is not due to chance. Then we can hypothesize that a common factor causes the relationship. In the example about height and weight, we might hypothesize that body mass is the common factor. As one grows taller, body mass, and hence weight, increases. It is the common factor of body mass, not chance, that causes the relationship between height and weight.

The opposite relationship is not necessarily true. It is certainly possible to gain weight without getting taller! A significant correlation does not prove causation; it only shows that a nonchance relationship exists. The fact that two variables are related does not mean that a change in one will necessarily produce a corresponding change in the other. Causation may be proved with other research techniques (such as comparing changes in an experimental group with changes in a control group after an independent variable has been allowed to operate on the experimental group) but not by correlation alone. Experimental design techniques to identify causation are presented in chapters 8 and 9.

When r is evaluated, the number of pairs of values from which the coefficient was calculated (N) is critical for determining the odds that the relationship could have happened by chance. If N is small, r must be large to be significant. When N is large, small r values may be significant. As we discussed earlier, low r values may indicate a real, or significant, relationship, but they usually are not useful for predicting individual scores.

Table A.2 in appendix A can be used to determine the significance of a correlation coefficient. First we calculate the **degrees of freedom** (df) for a correlation coefficient: $df = N_{pairs} - 2$, where N_{pairs} represents the number of pairs of XY scores. One degree of freedom is lost for each of the two variables in the correlation. The degrees of freedom represent the number of values that are free to vary when the sum of the variables is set.

Degrees of freedom compensate for small values of N by requiring higher absolute values of r before the coefficient can be considered significant. Notice in table A.2 that as the degrees of freedom increase, the absolute value of the coefficient needed to reach significance at a given level of confidence decreases. When N is large, the odds that a high r value may occur by chance are less than when N is small.

To read table A.2, find the degrees of freedom (df) in the left-hand column. Then proceed across the df row and compare the obtained r value with the value listed in each column. The heading at the top of each column indicates the odds of a chance occurrence, or the probability of error when declaring r to be significant. In the height and weight example from table 7.2, the degrees of freedom are calculated to be 8 ($N_{pairs} - 2 = 10 - 2 = 8$).

Table A.2 indicates that for $df = 8$, a correlation as high as .549 occurs only 10 in 100 times by chance alone ($p = .10$), an r value of .632 occurs only 5 times in 100 by chance ($p = .05$), and $r = .765$ occurs only 1 time in 100 by chance ($p = .01$). Previously we calculated an r value of .891 for the height and weight example. This value is greater than that reported in table A.2 for $p = .01$, so we can report with a better than 99% LOC that the value of $r = .891$ did not occur by chance. This result may be written as $r = .891$, $p < .01$. Use $p < .01$ rather than $p = .01$ because .891 is greater than the critical value from table A.2 of .765.

If the obtained r value is between the values in two columns, the left of the two columns (greater odds for chance) is chosen. If the obtained r does not equal or exceed the value in the .10 column, it is said to be not significant (N.S.). Negative r values are read in the same manner, but the absolute value of r is used.

When the r value is found to be significant, the cause of the correlation cannot be determined from the correlation data alone, although reasoned logic may point to a probable cause. As indicated earlier, further experimental evaluation is needed to determine direct causation in a correlation coefficient.

Bivariate Regression

When the correlation between two variables is sufficiently high, we can predict how an individual will score on variable Y if we know his or her score on variable X. This is particularly helpful if measurement is easy on variable X, but difficult on variable Y. For example, it is easy to measure the time it takes a person to run 1.5 miles and difficult to measure $\dot{V}O_2$max on a treadmill or bicycle ergometer. But if the correlation between these two variables is known, we can predict $\dot{V}O_2$max from 1.5-mile-run time. Of course, this prediction is not perfect; it contains some error. But we may be willing to accept the error to avoid the difficult and expensive direct measure of $\dot{V}O_2$max.

The following example does not involve data that are difficult to collect, but it does illustrate how an individual's score on one variable can be used to predict his or her score on a related variable. Suppose we gave a test of sit-ups per minute to a group of 15 high school boys on two consecutive days. Based on the data, we compute the following:

1st Day (X)	2nd Day (Y)
$\bar{X} = 52.33$	$\bar{Y} = 50.87$
$\sigma_X = 12.31$	$\sigma_Y = 10.56$

$$r_{XY} = .845.$$

Next we plot the values for each subject on an XY graph, as shown in figure 7.6. The 1st-day score is plotted on the X-axis and the 2nd-day score on the Y-axis. Note the example of a data point in figure 7.6 for a person who did 58 sit-ups the first day and 51 sit-ups the second day.

As we discussed earlier in this chapter, a scatter plot of the data clusters around the best fit line. When the best fit line is known, any value of X can be projected vertically to the line, then horizontally to the Y-axis, where the corresponding value for Y may be read.

The best fit line is graphically created by drawing it in such a way that it balances the data points. That is, the total vertical distance from each point below the line up to the line is balanced by the total vertical distance from each point above the line down to the line. The average distance of the points above the line is the same as the average distance of the points below the line. The vertical distance from any point to the line is called a **residual**. Residuals may be positive or negative.

Using the best fit line, we can predict the number of sit-ups that would be performed on the 2nd day (if a boy were absent and missed the 2nd test), based on the number he performed on the 1st day. We simply find the 1st-day score on the X-axis—for example, 58 sit-ups—and proceed vertically up to the line then horizontally over to the Y-axis to read the 2nd-day score. The 2nd-day score is about halfway between 50 and 60, so we estimate that the boy would have done 55 sit-ups on

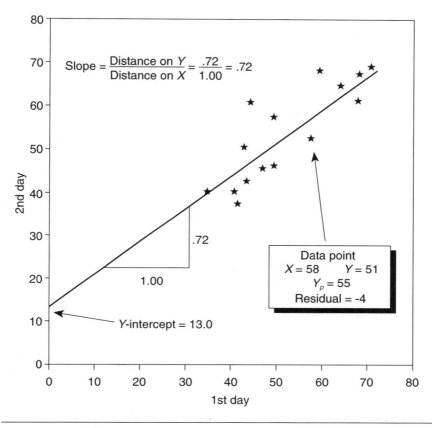

Figure 7.6 Prediction of Y from X on a scatter plot.

the 2nd day. This answer is not exact, because we must visually, or graphically, locate the point on the Y-axis.

The Y-coordinate can be determined more precisely by using the general algebraic formula for a straight line:

$$Y = bX + C, \qquad (7.04)$$

where Y is a value on the Y-axis, X is a value on the X-axis, b is the slope of the line, and C is the Y-intercept of the line.

The **Y-intercept** is the point where the extension of the best fit line intercepts the Y-axis. In figure 7.6, the Y-intercept is 13.0.

The slope is the tilt of the best fit line. In figure 7.6, the line advances only 72% as far on the Y-axis as it does on the X-axis when it progresses from left to right on the graph. This ratio, the distance traveled on the Y-axis divided by the distance traveled on the X-axis, is called the slope of the line. The slope is the tangent (opposite over adjacent) of the angle of intercept (the angle formed by the best fit line and a line parallel to the X-axis). In figure 7.6, the slope = +.72. From a table of trigometric values, we can determine that the tangent of +.72 is about 36 degrees. If the correlation coefficient is negative, the line tilts the other way, and the slope is negative.

When the means, standard deviations, and correlation coefficient are known, the slope and the Y-intercept are easily calculated. The formula for the slope (b) of the line is

$$b = \left[\frac{r(\sigma_Y)}{\sigma_X} \right],$$ (7.05)

and the formula for the Y-intercept (C) is

$$C = -\left[\frac{r(\sigma_Y)}{\sigma_X} \right] \bar{X} + \bar{Y}.$$ (7.06)

When we substitute these expressions for b and C into the general formula for a straight line, $Y = bX + C$, the expanded equation for Y_p becomes

$$Y_P = \left[\frac{r(\sigma_Y)}{\sigma_X} \right] X - \left[\frac{r(\sigma_Y)}{\sigma_X} \right] \bar{X} + \bar{Y},$$ (7.07)

where Y_p is the predicted value of Y given X. Using equation 7.05, we solve for slope:

$$b = \left[\frac{.845\ (10.56)}{12.31} \right] = .72.$$

Then we use equation 7.06 to find the Y-intercept:

$$c = -\left[\frac{.845\ (10.56)}{12.31} \right] 52.33 + 50.87 = 13.0.$$

The generalized equation for Y_p becomes

$$Y_p = .72\ (X) + 13.0.$$

If X = 58, then

$$Y_p = .72\ (58) + 13.0 = 54.76.$$

This value for Y_p (54.76) when $X = 58$ is reasonably close to the value that we graphically predicted earlier (55), but the algebraic solution is more accurate. Now that the generalized equation $Y_p = .72 (X) + 13.0$ is known, we can substitute any value for X and easily predict its concomitant Y value.

This statistical process is called **bivariate regression** analysis. Regression was first observed in 1877 by Sir Francis Galton (1822–1911) when he noted that characteristics of offspring tended to cluster between the characteristics of their parents and the mean of the population (Kotz & Johnson, 1982, p. 275). In other words, the characteristics of the offspring tended to "regress" toward the mean. By using this technique, scientists could predict the characteristics of the offspring. In application today, the word prediction is probably more descriptive of the process than regression. When you hear regression, think prediction.

Regression is useful for predicting scores on variables that are difficult to measure, such as $\dot{V}O_2max$. But as was mentioned earlier, this technique involves some error. How large is the error in the prediction?

Determining Error in Prediction

To demonstrate how to calculate the error in prediction we'll use the sit-up example from figure 7.6. Each data point has an error factor, the distance from the best fit line, called a residual. These distances represent the residual, or the part left over, between each predicted Y value and the actual value. Sometimes they are called residual errors because they represent the error in the prediction of each Y value. The boy who did 58 sit-ups on the 1st day would be predicted (see figure 7.6) to do about 55 sit-ups (Y_p) on the 2nd day. If he actually did 51 (Y), the residual error would be –4, the difference between the actual ($Y = 51$) and the predicted ($Y_p = 55$).

The best fit line represents the best prediction of Y for any X value. Some residuals are large, and some are quite small; indeed some points fall very close to the line. By using the algebraic solution for the best fit line and the residuals, we can calculate the predicted value for Y and the amount of error in the prediction.

For example, what is the prediction of Y and the error of prediction for a different subject (not one of the original 15) who did 60 sit-ups on the 1st day, but was absent on the 2nd day?

The generalized equation $Y_p = .72 (60) + 13.0 = 56.2$ produces the predicted Y value, 56.2, but this prediction has some degree of error. To determine the error in prediction, we could use the generalized equation to predict a Y value for each subject and then compare the predicted Y value (Y_p) with the actual Y value to determine the amount of error in prediction for each subject ($Y - Y_p$). The result of such an analysis is presented in table 7.3.

The values for $Y - Y_p$ represent all the errors, or residuals, around the mean of the best fit line. We can assume that they are randomly distributed around the line and that they would fall into a normal curve when plotted. In chapter 5, we defined

Table 7.3   Calculation of the Standard Error of the Estimate

X	Y	Y_p	$Y–Y_p$	$(Y–Y_p)^2$
72	68	65.03	2.97	8.82
70	64	63.59	.41	.17
69	58	62.87	–4.87	23.72
67	61	61.43	–.43	.18
61	62	57.11	4.89	23.91
58	51	54.95	–3.95	15.60
51	55	49.91	5.09	25.91
49	44	48.47	–4.47	19.98
47	43	47.03	–4.03	16.24
43	60	44.15	15.85	251.22
42	46	43.43	2.57	6.60
41	38	42.71	–4.71	22.18
40	36	41.99	–5.99	35.88
40	37	41.99	–4.99	24.90
35	40	38.39	1.61	2.59
				$\Sigma(Y–Y_p)^2 = 477.90$

standard deviation as the square root of the average of the squared deviations from the mean. Therefore, the standard deviation of the residuals can be calculated with a modification of equation 5.03 as follows:

$$\sigma_{res} = \sqrt{\frac{\Sigma(Y - Y_p)^2}{N}}. \tag{7.08}$$

Then

$$\sigma_{res} = \sqrt{\frac{477.90}{15}} = 5.64.$$

Equation 7.08 is generally not used because of the tedious calculations involved. Another formula, the standard error of the estimate (SE_E) formula, is much easier to use. The SE_E formula requires only the standard deviation of the Y variable and r_{XY}:

$$SE_E = \sigma_Y \sqrt{1 - r^2}. \tag{7.09}$$

This formula (sometimes denoted by σ_{est}) yields the same answer as equation 7.08:

$$SE_E = 10.56 \sqrt{1 - (.845)^2} = 5.64.$$

The standard error of the estimate may be interpreted as the standard deviation of all the errors, or residuals, made when predicting Y from X. Because it is the standard deviation of a set of scores (the residuals) that are normally distributed, the SE_E can be interpreted as a Z score of ± 1.0. We know that 68% of all errors of prediction will be between $\pm 1 \times SE_E$, 90% will be between $\pm 1.65 \times SE_E$, 95% will be between $\pm 1.96 \times SE_E$, and 99% will fall between $\pm 2.58 \times SE_E$. (See table 6.1 for a review of the relationships between Z, LOC, and p.)

We are now prepared to estimate error for the predicted Y value for a subject who performed 60 sit-ups on the 1st day but was absent on the 2nd. The generalized equation predicts the Y value to be

$$Y_p = .72(60) + 13.0 = 56.2 \text{ sit-ups.}$$

The SE_E (from equation 7.09) in this prediction is ± 5.64 at the 68% level of confidence. This means the prediction of $56.2 \pm 1.0 \times 5.64$ sit-ups has a 32% chance of being incorrect. When the Z score is 1.96, the error in the prediction drops to 5% ($1.96 \times 5.64 = 11.1$). The prediction then becomes

$$Y_p = 56.2 \pm 1.96 \ (5.64) = 56.2 \pm 11.1, \quad p = .05.$$

In other words, the odds are 95 to 5 that the predicted value of Y lies between 45.1 and 67.3, or in integer values, 45 and 67. The final form of the generalized regression equation is represented below:

$$Y_p = b\ (X) + C \pm Z\ (SE_E), \tag{7.10}$$

where Z is a value that produces the desired level of confidence.

Once this formula is established for a given problem, predictions with error estimations can be made from any value of X when the means, standard deviations, and correlation are known. The critical factors are the standard deviation of Y (s_y) and r. When s_y is small and r is large, the errors in prediction are small. When r drops below about .7, the errors may become so large that the prediction may not be useful. In the example described above, if $r = .7$ rather than .845, the error increases:

$$SE_E = 10.55 \ \sqrt{1 - .7^2} = 7.53.$$

And if r is .5, then

$$SE_E = 10.55 \ \sqrt{1 - .5^2} = 9.14.$$

Each researcher must determine an acceptable error level for the data being analyzed. That level will depend on the consequences of erroneous predictions.

Multiple Regression

The bivariate regression concepts described previously can be expanded to apply to **multivariate** data. Such data include three or more variables where one variable—the dependent variable, usually labeled Y—is related to and dependent on two or more independent variables—usually labeled X_1, X_2, X_3, and so on. Multiple regression relationships are represented by R to differentiate them from simple bivariate values symbolized by r. The advantages of **multiple regression** over bivariate regression are (a) multiple regression usually provides a lower standard error of the estimate, and (b) it provides us with information to determine which independent variables contribute to the prediction and which do not. The general formula for multiple regression is represented by:

$$Y_p = b_1 X_1 + b_2 X_2 + b_3 X_3 \ldots + b_k X_k + C, \qquad (7.11)$$

where $b_1, b_2, b_3, \ldots, b_k$ are coefficients that give weight to the independent variables according to their relative contributions to the prediction of Y. The number of predictor, or independent, variables is represented by k, and C is a constant; it is similar to the Y-intercept. When raw data are used, the b values are in raw score units. Sometimes multiple regression is performed on standardized scores (Z) rather than on raw data. In such cases, all raw scores are converted to the same scale, and the b values are then referred to as **beta weights.** Beta weights perform the same function as b values; that is, they give relative weight to the independent variables in the prediction of Y. In common usage, b values are sometimes called beta weights, but the word beta is only properly used when the equation is in a Z score format.

Multiple regression is used to find the most satisfactory solution to the prediction of Y, which is the solution that produces the lowest standard error of the estimate (SE_E). Each predictor variable is weighted so that the b values maximize the influence of each predictor variable in the overall equation.

Multiple regression has been used in exercise physiology to determine the best combination of skinfold sites for predicting body composition. First, an accurate measure of body density (the dependent variable, Y) is taken (usually hydrostatically, but other even more accurate methods may be used). Multiple skinfold sites are then measured with calipers, each site representing an independent variable (X_1, X_2, X_3, etc.). When the multiple regression equation is determined, the most predictive of the skinfold sites are used to determine body density.

Standard multiple regression results in one equation with all appropriate independent variables and the constant included. However, there may be occasions when we want to force certain independent variables into the equation before others.

Hierarchical multiple regression is the process where we override the computer and set up a hierarchical order for inclusion of the independent variables. This may be useful if some of the independent variables are easier to measure than others, or if some are more acceptable to use than others.

A third strategy is called **stepwise multiple regression.** In this process, the computer produces a series of equations, first a bivariate solution, then additional equations in a step-by-step order as other independent variables enter the multiple regression solution. The final stepwise equation will be the same as the single equation produced by standard multiple regression if the same set of predictor variables is used. Stepwise regression offers the advantage of listing the order of the steps in the development of the equation so that we can identify the effect of each variable on the SE_E as it enters the equation.

Multiple regression is a complicated statistical process; it will not be mathematically demonstrated here. A detailed analysis of multiple regression may be found in advanced statistical texts; Tabachnick and Fidell (1989, chapter 5) present a particularly good analysis that includes theoretical, practical, and computer applications. Computer programs such as SPSS (Statistical Package for the Social Sciences), BMDP (Biomedical Programs), and SAS (Statistical Analysis Systems) can perform multiple regression analyses.

Figure 7.7, modified from Tabachnick and Fidell (1989, p. 142), uses a Venn diagram to demonstrate a theoretical explanation of how independent variable influence in multiple regression is determined. The dependent variable (DV) is represented by Y, and X_1, X_2, X_3, and X_4 are independent variables (IVs). Variable X_4 is not related, or has no common variance, with Y so it does not enter into the equation. Variable X_2 has the greatest common variance with Y (represented by areas B, C, and D), so it contributes the most to the equation and is entered first in the stepwise process. Variable X_1 contributes less than X_2, but more than X_3; the unique variance of X_1 (represented by area A) is smaller than the unique variance of X_2 (area C) but larger than the unique variance of X_3 (area E). Variable X_3 contributes least, or perhaps not at all if the addition of area E does not significantly improve the prediction made by X_1 and X_2 (areas A, B, C, and D).

Partial correlation is the correlation of a subsequent IV with the DV after the variance in the DV, which is explained by a previous IV, has been removed. For example, if X_2 explains areas B,C, and D in the variance of Y, then the partial correlation (also called unique variance) of X_1 with Y is area A.

The independent variables X_1 and X_2 are related to each other (they have common variance in areas B and F). Variables X_2 and X_3 also possess common variance (areas D and G), but X_1 and X_3 are not related. Because some of the independent variables are related to each other, only the one with the largest common variance with Y (in this case, X_2) is used totally in the equation. After X_2 has entered the equation, X_1 can add only area A, and X_3 can add only area E to the prediction. It is common for independent variables to be related to each other, and this may reduce their effectiveness in adding to the prediction because each can contribute only the unique variance that it possesses.

Figure 7.8 represents the ideal relationship arrangement for dependent and independent variables. The independent variables—X_1, X_2, X_3, and X_4—are all highly related to Y but are not highly related to each other. Each independent variable may make a large unique contribution to the prediction; together the

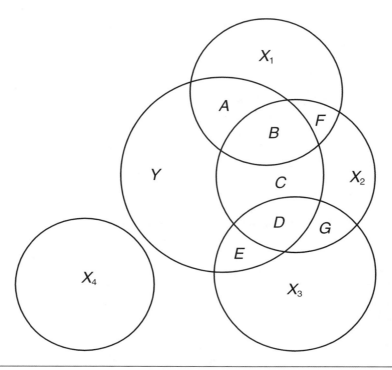

Figure 7.7   Venn diagram of unique and common variance in multiple regression.

independent variables explain almost all of the variance in Y. It is not common, however, to find this arrangement in practical research. Usually the independent variables are related to each other as well as to Y, thus reducing the effectiveness of some of them in adding to the prediction.

When all independent variables are highly related to the dependent variable (Y) but not to each other, the result is a small SE_E and consequently a good prediction equation. As in figure 7.8, almost all of the variance in Y can be explained by the independent variables. To produce the multiple regression equation that has the highest predictive value and the lowest SE_E, investigators should seek independent variables that are highly related to the dependent variable but are not related to each other.

It is possible for two independent variables to be both highly related to the dependent variable and highly related to each other. Figure 7.8 depicts a case in which two independent variables, X_3 and X_5, both overlap Y but also lie almost on top of one another. Of the two variables, only X_3 will enter the equation, because once X_3 is entered, X_5 will have no unique variance with Y and will not influence the prediction equation.

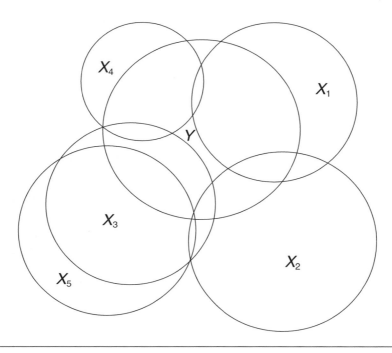

Figure 7.8   Ideal relationships among dependent and independent variables in multiple regression.

An Example From Exercise Physiology

Equations to predict body composition are population specific: An equation developed on one subset of the population may not be accurate for another subset. Variations in age, gender, ethnicity, fitness levels, nutrition, and other factors influence the equation. If we needed an equation to predict percent body fat in sedentary midlife females, and one was not readily available in the literature, we would have to develop our own specific equation. To do this, we might randomly select 200 sedentary female subjects between the ages of 40 and 60 and determine their percent body fat hydrostatically. Then we would take skinfold measurements in millimeters at multiple sites, such as triceps, subscapular, abdominal, suprailiac, and calf.

Using hydrostatically determined percent body fat as the dependent variable, and measurements for the five skinfold sites as independent variables, we might enter the data for the first 100 of the subjects into a computer and receive the following printout for standard multiple regression:

R = .868

Constant = 5.26

Standard error (SE_E) = 3.67

R^2 = .753

Dependent	Mean	SD
Percent body fat (Pfat)	22.4	7.39

Variables in	b	Mean	SD	Variables not in
Abdominal (Abd)	.712	21.88	4.66	Suprailiac*
Triceps (Tri)	.224	12.56	3.21	Subscapular*
Calf	.187	13.95	4.17	

*These variables are not included in the equation because they did not add a sufficient unique variance to make a significant difference in the accuracy of the prediction after the first three variables were entered.

Using these data, we begin with the general equation

$$Y_p = b_1 (X_1) + b_2 (X_2) + b_3 (X_3) + C \pm (SE_E).$$

Then we substitute b values and the constant from the computer printout. The general equation for Pfat becomes

$$\text{Pfat} = .712 \ (\text{Abd}) + .224 \ (\text{Tri}) + .187 \ (\text{Calf}) + 5.26 \pm (3.67).$$

For a female whose skinfold site measurements in millimeters are abdominal = 20, triceps = 15, and calf = 17, we would calculate her percent body fat to be

$$\text{Pfat} = .712 \ (20) + .224 \ (15) + .187 \ (17) + 5.26 \pm (3.67).$$

or

$$\text{Pfat} = 20.80 \pm 3.67.$$

This calculation indicates that the best estimate of the subject's percent body fat is 20.80 with a potential error of ±3.67. Because SE_E represents ±1 Z on a normal curve, the odds are better than 2 to 1 (68% LOC) that the subject's true percent body fat lies somewhere between 22.37 and 29.71. For higher levels of confidence, the SE_E could be multiplied by an appropriate Z value (1.65 for p = .10, 1.96 for p = .05, or 2.58 for p = .01).

Next we apply the general formula for Pfat to each of the other 100 subjects in the data set and compare their actual percent body fat (measured hydrostatically) with their percent body fat predicted by the equation. This process of developing the equation on one set of subjects and testing the prediction on another set is called **cross validation.** The predicted body fat values on the second set of 100 subjects are statistically compared with their actual measured values to assess the accuracy of the prediction.

Some Cautions and Assumptions

When performing a multiple regression analysis, be aware of the following assumptions and how they may affect your results.

Ratio of Subjects to Independent Variables

The ratio of subjects to independent variables should be no less than 5:1, and ideally about 20:1. In stepwise regression, an even greater ratio should be sought (40:1). Reducing this ratio seriously limits the ability to generalize the equation. However, it is also possible to have too many cases. With a large enough N, almost any multiple correlation may be found to be significant. For most studies, ratios between 20:1 and 40:1 are reasonable (Tabachnick & Fidell, 1989, p. 129).

Outliers

Outliers, cases with excessively large residual values, can produce greater leverage on the resultant equation than is appropriate. They should be eliminated from the database before analysis. Multivariate outliers are especially devious because they are hard to find. It may not be too unusual for a person to be 4'6" tall or to weigh 250 pounds; such values are moderate univariate outliers. But a person who is 4'6" tall and also weighs 250 pounds is clearly not typical of the population and represents an extreme multivariate outlier.

Outliers may represent real but extreme cases in the population, or they may be the result of measurement errors introduced during data collection or data entry. Outliers caused by errors must be found and corrected. Real, extreme case outliers must be evaluated by the researcher to determine if they are truly representative of the population to be studied.

L.S. Fidell (personal communication, 1992) stated that:

> The problem with outliers is that they influence the solution too much; therefore, the solution does not generalize well to normal populations. An outlier may, in fact, have a "smaller than it should have had" residual because it pulled the solution towards itself, thus biasing the prediction equation.

Advanced computer programs such as SPSS, BMDP, and SAS can test for both univariate and multivariate outliers.

Normality and Homoscedasticity of Residuals

The residuals for each value of X are assumed to be normally distributed around the best fit line, and the variances of each set of residuals (X_1, X_2, X_3, etc.) must be approximately equal. The condition of equal residual variance is called **homoscedasticity.** Both Kachigan (1986) and Tabachnick and Fidell (1989) discuss this condition in more detail. Software programs such as SPSS, BMDP, and SAS can be used to test for this assumption.

Multicollinearity and Singularity

Multicollinearity means that two or more of the independent variables are highly correlated. Recall that ideally the independent variables should be related to the dependent variable but not to each other (see figures 7.7 and 7.8). Variables highly related to a variable already entered into the equation usually do not add to the prediction. In figure 7.8, X_3 and X_5 are multicollinear.

Singularity means two or more independent variables are perfectly related to each other ($r = 1.00$). This may occur if one variable is created from another by a mathematical manipulation such as squaring, taking the square root, or adding, subtracting, multiplying, or dividing by a constant. Most advanced computer programs screen for multicollinearity and singularity and warn the user in the printout if these relationships are detected by producing the squared multiple correlation (SMC) values for all variables. See Tabachnick and Fidell (1989, p. 87) for further discussion of this issue.

Cross Validation

Sound research design requires that the results be tested for accuracy. In the development of either bivariate or multivariate regression equations, the equation should be developed on one sample of the population and then tested on another equivalent sample. This may require dividing the sample in half, which would reduce the size of N. But without cross validation on an equivalent sample to test the accuracy of the prediction, the results will be suspect.

Summary

Correlation (either bivariate or multivariate) is designed to determine the relationships between or among variables. Correlation requires measurements of at least two variables on the same set of subjects. Pearson's correlation coefficient evaluates the relationship by comparing the Z score deviations from the means for the two variables on each subject. If the correlation is high and positive, a person is likely to score high on both variables, or low on both variables, or in the middle on both variables. If the correlation is high and negative, a person will tend to score high on one variable and low on the other. The absolute value of the coefficient can be evaluated by using table A.2 in appendix A to determine the probability that the correlation is reliably different from 0.00—that is, the probability that the correlation did not happen by chance. The significance of difference between the r value and an assumed population value can also be determined by a special form of the t test (Witte, 1985, p. 180).

Correlation is often used to predict one variable from the value of another. This ability to predict is most useful when it is difficult to measure one of the variables. The prediction always has some error in it, unless the correlation is +1.00 or –1.00.

When deciding to use prediction, we must determine if it is more acceptable to tolerate the error in the prediction or the difficulty of direct measurement.

Correlation is useful in determining the quality of measurement. Validity can be measured with correlation by determining the relationship between a known valid test and a test of unknown validity. If the tests correlate highly, then they are measuring the same factor or factors. Reliability, the consistency of the measurement assessed by intraclass correlation (see chapter 10), is an essential component of quality measurement. Objectivity, a measure of bias in the human judgment of performance, can also be assessed with intraclass correlation. Intercorrelations among the ratings of judges are one way to identify bias (either conscious or unconscious) among judges. Correlation is widely used in kinesiology.

Problems to Solve

1. Estimate whether the correlation between the following pairs of variables is positive, negative, or zero:
   A. Height and weight
   B. Upper body strength and distance in the shot put
   C. Arm length and softball throw for distance
   D. $\dot{V}O_2$max and place in a cross-country meet
   E. Standing long jump and vertical jump test score
   F. Swimming sprint speed and tennis serving accuracy
   G. Golf score and the number of hours spent practicing
   H. School grade point average and hours spent watching television
   I. Daily temperature and the average weight of clothing worn
   J. $\dot{V}O_2$max and 2-mile-run time

2. Calculate r from the following data ($N = 10$):

	$\bar{X}$	$\bar{Y}$	σ_X	σ_Y	ΣXY
A.	12	15	3	4	1,900
B.	12	15	3	4	1,700

3. Use table A.2 in appendix A to determine the p value for the following:

	r	N
A.	.600	13
B.	.600	20
C.	.600	8

4. A physical education department wanted a single test of upper body strength that was easy to administer. Dips on the parallel bars and pull-ups on the

horizontal bar were considered good tests. One faculty member thought that both tests were not needed, because the correlation between the two was probably high. To evaluate this assumption, 141 students were tested on both criteria. The faculty member let X represent dips on the parallel bars and Y represent pull-ups and calculated the following from the data:

$$\Sigma X = 3{,}416,\ \Sigma Y = 1{,}899,\ \Sigma X^2 = 93{,}810,\ \Sigma Y^2 = 28{,}697,\ \Sigma XY = 50{,}509.$$

Calculate the following:

A. The mean of each variable
B. The standard deviation of each variable
C. The correlation between the two variables. Use the mean/standard deviation formula and the machine formula. Do they agree?
D. The level of confidence and the p value reached by the coefficient
E. The predicted number of pull-ups for a student who performed 20 dips
F. The standard error of the estimate (SE_E)
G. The predicted range of possible pull-up scores at the 95% level of confidence for a student who performed 20 dips

5. In a motor learning lab, data on time in milliseconds for a simple reaction time (RT) and movement time for a linear arm movement (MT) were collected on 10 subjects. Is there a relationship between RT and MT? If you have a computer program available that computes correlation, use it on this problem. (Data from California State University Northridge motor learning lab, courtesy of Tami Abourezk.)

RT	MT
271	354
268	435
198	211
345	411
169	288
209	413
322	158
209	333
199	425
216	378

6. A kinesiology major wanted to predict $\dot{V}O_2$max based on the mile run. To develop the regression equation, she obtained $\dot{V}O_2$max values (ml/kg/min) in the exercise physiology laboratory on 18 students. Two days later, she measured the same 18 students on the mile walk/run with scores reported as total time in seconds. Data follow.

Subject	Mile Walk/Run	$\dot{V}O_2$max
1	250	60.3
2	315	57.2
3	420	55.4
4	410	51.4
5	436	52.5
6	511	45.6
7	460	38.4
8	510	41.5
9	530	39.6
10	586	33.2
11	591	37.7
12	600	40.1
13	626	32.0
14	643	35.4
15	650	33.7
16	675	35.9
17	710	27.4
18	720	25.3

A. Using a computer, calculate the means, standard deviation, and correlation coefficient between the two variables. Then, using only a calculator, confirm your answer by hand.

B. What is the probability that the correlation coefficient happened by chance? Why is the coefficient negative?

C. What is the slope, the Y-intercept, and the standard error or the estimate for the regression line?

D. Estimate $\dot{V}O_2$max for a person who runs the mile in 540 seconds. If you wanted your estimate to be accurate at the 95% level of confidence, what is the error factor in your estimate?

7. Measure the height in inches and the weight in pounds of about 20 of your classmates.

A. What is the correlation between the two variables? First calculate the answer by hand. Then, if a computer is available, enter the data and compare your hand-calculated answer to the computer's answer.

B. Is the correlation significant?

C. If you concluded that it is significant, what is the probability of error in your conclusion?

See appendix C for answers to problems.

Key Words in This Chapter

Correlation
Pearson's product moment
    correlation coefficient
Coefficient
Scatter plot
Best fit line
Curvilinear
Degrees of freedom
Y-intercept
Bivariate regression
Multivariate

Multiple regression
Beta weights
Standard multiple regression
Hierarchical multiple regression
Stepwise multiple regression
Partial correlation
Cross validation
Homoscedasticity
Multicollinearity
Singularity

Chapter 8

The *t* Test: Comparing Means From Two Sets of Data

If we want to know whether the mean resting heart rate of cross-country runners is different from that of gymnasts at a given university, we can simply measure resting heart rates of all the members of the cross-country team and all the members of the gymnastics team and compare the results. If the runners average 50 beats per minute and the gymnasts 60, it is clear that resting heart rate is lower for the runners—no statistical analysis is needed. However, if we want to compare *all* collegiate runners with *all* collegiate gymnasts, then we are faced with the task of measuring a great number of athletes. An easier alternative is to sample each population and compare the means of the samples. We can then determine whether the mean differences found in the samples can be reliably applied to the populations from which the samples were drawn. The technique is called a *t* test. In this chapter, we will learn the method for doing this.

Recall from chapter 6 that a sample mean may be used as an estimate of a population mean. Remember also that we can determine the odds, or probability, that the population mean lies within certain numerical limits by using the sample mean as a predictor. This same technique can be used in reverse to determine if a given sample is likely to have been randomly selected from a specific population.

If the population mean is known or assumed to be a certain value, and if the sample mean is not close enough to the population mean to fall within the limits set by a selected level of confidence, then one of the following conclusions must be true: (a) The sample was not randomly drawn from the population, or (b) the sample was drawn from the population, but it has been modified so that it is no longer representative of the population from which it was originally drawn.

Using similar logic, we can make conclusions about two sets of data. If we draw two samples from the same population, and the means of these samples differ by amounts larger than would be expected based on normal distributions, one of the following conclusions must be true: (a) One or both of the samples were not randomly drawn from the population, or (b) some factor has affected one or both samples, causing them to deviate from the population from which they were originally drawn.

t Tests

When a sample is drawn from a population with a known or estimated mean (μ) and standard deviation (σ), the probability (or odds) that the mean of a randomly drawn sample (X) will lie within certain limits of μ can be determined. To ascertain the probability that a given sample came from a certain population, the value of the standard error of the mean must be calculated by one of the following formulas. If the standard deviation of the population (σ) is known, the formula used is

$$\sigma_M = \frac{\sigma}{\sqrt{N}}, \qquad (8.01)$$

where σ_M is the symbol for the actual standard error of the mean for a population with known mean and standard deviation. If σ is not known, the formula used is

$$SE_M = \frac{SD}{\sqrt{N}}, \qquad (8.02)$$

where SE_M is the standard error of the mean estimated from the sample. Note that this is the same as equation 6.03.

Using either σ_M or SE_M, we can determine the odds that a sample is representative of the population from which it was drawn by doing a Z test, if the population is known,

$$Z = \frac{\overline{X} - \mu}{\sigma_M},$$ (8.03)

or a *t* test,

$$t = \frac{\overline{X} - \mu}{SE_M},$$ (8.04)

if the population is estimated.

Evaluating Z

When σ_M is known, Z indicates the significance of the difference between $\overline{X}$ and μ. Under these conditions, we can determine the significance of Z by comparing it to critical ratios of 1.65 at $p = .10$, 1.96 at $p = .05$, and 2.58 at $p = .01$. If Z exceeds one of these levels, we may conclude at the given level of confidence (LOC) that $\overline{X}$ was not randomly drawn from μ.

Evaluating t From a Single Sample

The approximation of σ_M by SE_M is not accurate in small samples (N less than 60). This was first demonstrated by an English statistician named William Sealy Gossett (1876–1937), who wrote under the pseudonym Student (Kotz & Johnson, 1982). He developed a series of approximations of the normal curve to account for the bias in the estimate of σ_M called *Student's t distribution*. Table A.3 in appendix A lists the values for Student's *t* distribution. If there were no error in the estimation, table A.1 could be used in every case. But when samples are used to estimate population parameters, especially when the samples are small, the *t* distribution (table A.3) must be used to evaluate the *t* statistic. The values in table A.3 at the given *p* levels are called **critical ratios.** They represent the *t* ratio that must be reached to reject chance.

The *t* test for one sample produces the ratio of the **actual mean difference** between the sample and the population to the **expected mean difference,** or that amount of difference between $\overline{X}$ and μ that can be expected to occur by chance alone. The expected mean difference is estimated by equation 8.02 and is called the standard error of the mean. To interpret *t* for a single sample, we must first find the degrees of freedom, which can be calculated by the formula $df = N - 1$. The *t* ratio is compared to the values for a two-tailed test (one- and two-tailed tests will be explained later in this chapter) from the *t* distribution in table A.3 for the appropriate df.

When *t* exceeds the value in table A.3 for a given *p* level, we may conclude that $\overline{X}$ was not drawn from μ. When *t* is less than the critical ratio in table A.3, the null hypothesis (H_0) is accepted; there is no reliable difference between $\overline{X}$ and μ. When *t* exceeds the critical ratio, H_0 is rejected and H_1 is accepted; some factor other than chance is operating on the sample mean. Notice that in table A.3 when degrees of freedom are large ($df > 120$), the values for a two-tailed t test at a given p value are the same as the values read from table A.1 for a Z test (1.65, 1.96, and 2.58).

This technique is useful for determining if influences introduced by an experiment have an effect on the subjects. If we know or estimate the population parameters and then draw a random sample and treat it in a manner that is expected to alter its mean value, we can determine the odds that the treatment had an effect by using a t test. If t exceeds the critical ratios in table A.3, we can conclude that the treatment was effective because the odds are high that the sample is no longer representative of the population from which it was drawn. The treatment has caused the sample to change so that it does not match the characteristics of the parent population.

Assumptions for the t Test

Several assumptions must be met for the t test to be properly applied. If these assumptions are not met, the results may not be valid. When the investigator knows that one or more of these criteria are not met, a more conservative (i.e., $p = .01$ rather than $p = .05$) level should be selected to avoid errors. This allows us to be confident of the conclusions and helps to compensate for the fact that all assumptions were not met. The t test is quite robust; it produces reasonably reliable results, even if the assumptions are not met totally. The t test is based on the following assumptions:

- The population from which the samples are drawn is normally distributed. (See chapter 6 for methods of determining the amount of skewness in a data set.)
- The sample or samples are randomly selected from the population. If the samples are not randomly selected, a generalization from the sample to the population cannot be made.
- When two samples are drawn, the samples have approximately equal variance. The variance of one group should not be more than twice as large as the variance of the other. This is called homogeneity of variance.
- The data must be parametric, that is, based on an interval or ratio measurement scale (see chapter 1).

An Example From Physical Education

The faculty of a physical education department became concerned about the apparent lack of skill students developed in their 5-week volleyball units. Students are assigned to instructors in a random fashion alphabetically, based on last name. Typically the classes just play during the entire period and receive little or no practice on specific skills. A standardized test of volleyball serving ability (50 points, maximum) has been given to every student for many years and the data have been saved. The mean for more than 1,000 students (the population) on this test is 31 points (μ) with a standard deviation of 7.5 (σ).

One teacher decided to try a different approach. In one class (the sample), half the period was devoted to teaching skills, especially serving skills, and the students practiced under the teacher's direction for 20 minutes. Games were played only at the end of the period, and a tournament among the squads was held the last week of the volleyball block.

The students in this class ($N = 30$) were also given the standardized serving test. Their average score ($\bar{X}$) was 35 points out of the 50 possible with a standard deviation (*SD*) of 8.3. Was the teacher effective in improving serving skills? In other words, if we assume the class of 30 students to be a random sample of the population of more than 1,000 students, is it likely that the average score for the class ($\bar{X} = 35$) is representative of the population mean ($\mu = 31$)? To test the hypothesis that the sample class represents the population, we calculate the standard error of mean for the population:

$$\sigma_M = \frac{\sigma}{\sqrt{N}} = \frac{7.5}{\sqrt{30}} = 1.37.$$

Then we conduct a *Z* test (because μ and σ are known) to determine the odds that the mean of a sample randomly drawn from the population would differ from the population mean by as much as 4 points:

$$Z = \frac{35 - 31}{1.37} = 2.92.$$

What are the odds that a *Z* score of 2.92 would be found if the sample were drawn from the population and not treated? Because *Z* is greater than 2.58, the probability that $\bar{X}$ did not come from μ is greater than 99 to 1, $p < .01$.

Two possibilities must be considered:

1. The class was not a random sample of the population and therefore does not represent the population. Perhaps by luck, or by design, these 30 students were better at the beginning of the 5-week block than the typical students assigned to the other classes.
2. The sample was random at the beginning of the 5-week block, but the treatment (instruction and practice) has changed the students in such a way that they are no longer representative of the population. In other words, the students have changed so that they now represent another population, one that has instruction and practice rather than free play.

If random assignment to instructors can be demonstrated, so that there is assurance that the class was representative at the start of the 5-week block, then only one conclusion is left. A difference of 4 points between the mean of the sample and the mean of the population would occur less than 1 time in 100 by chance alone. In other words, the odds that the instruction was effective are better than 99 to 1 (LOC = 99%). We reject H_0 and conclude that the instruction was effective at $p < .01$.

In the previous example, we used σ because we knew its value. But in most research, μ and σ are not known. If σ is not known, we estimate SE_M, the standard error of the mean, using equation 8.02 and the standard deviation of the sample. Then we use the t test rather than Z to determine the significance of the difference between the sample and the *estimated population mean* (using equation 8.02). First we calculate SE_M:

$$SE_M = \frac{8.3}{\sqrt{30}} = 1.52.$$

Then we use SE_M to determine t. If $F_u = 35$, then

$$t = \frac{35 - 31}{1.52} = 2.63.$$

Notice that the answer for t is slightly smaller than the Z score (2.92). This demonstrates that the power to detect differences between samples and populations is greater when the population is known than when it is estimated. To find the level of confidence, we compare the t ratio (2.63) to the critical ratios of a two-tailed test from the t distribution in table A.3 for $df = 30 - 1 = 29$. The critical ratio at $df = 29$ in table A.3 for $p = .05$ is 2.045 and for $p = .01$ it is 2.756. Our t value falls between these values, so we accept the less significant value, $p < .05$. Note that we have a lower level of confidence (95%) for rejecting H_0 when the population is estimated than when the population parameters are known (99%).

Comparing Two Independent Samples (A Between Comparison)

The same concepts used to compare one sample mean to the population mean may be applied to compare two samples drawn from the same population. We will review the theory before considering the calculations. If the ratio exceeds the critical ratio from table A.3, the null hypothesis H_0 is rejected. In this example, H_0 states that both means represent (or were randomly drawn from) the same population. If the t ratio does not exceed the critical ratio, H_0 is accepted.

In this design, two independent samples are randomly drawn from a population. By **independent samples,** we mean that the subjects in one sample are different people and are not related, or correlated, in any way to the subjects in the other sample. One sample, the experimental group, is treated, and the other, the control group, is not. A *t* test is performed to compare the actual difference between the means of the two samples (the numerator of the *t* test) with the difference expected if only chance were operating (the denominator of the *t* test). This ratio is used to determine whether the two groups both represent the population from which they were drawn. If the calculated *t* ratio is larger than the critical ratio from the *t* distribution in table A.3, chance is rejected, and a probable cause is assumed. When H_0 is rejected for an experiment conducted under carefully controlled conditions, the treatment is identified as the probable cause. This design is represented in table 8.1 where two different groups of subjects each are compared before and after treatment.

If both groups are randomly drawn from the same population, a *t* test between the groups' means on the pretest should not be significant. If the treatment is effective, and all other variables are controlled, a *t* test between the groups on the posttest should be significant. This indicates that the experimental group is no longer representative of the original population from which it was drawn because the treatment has caused it to change.

In chapter 6 it was demonstrated that the means of a large number of random samples drawn from the same population will form a normal curve. The central limit theorem indicates this is true even if the population is skewed. **The differences between pairs of random samples drawn from a population will also form a normal curve.** If enough pairs of sample means are randomly drawn, and the difference between each pair is consistently calculated by subtracting the second mean from the first, then the average difference of all the pairs will be zero and the distribution will be normal. In some cases, the first mean will be larger than the second, making a positive difference; in other cases, the second will be larger than the first, producing a negative difference. When added together, the positives should cancel out the negatives, and all the difference scores should sum to zero.

If we assume an infinite set of scores, each derived from the mean difference of a pair of randomly drawn samples, these differences will form a normal curve with a mean of zero. After calculating the standard deviation of the difference scores (called the *standard error of the difference, SE_D*), we could then compute *t* scores and look in table A.3 to determine significance. Note that in this case, SE_D becomes the denominator of the *t* test. It is the value that indicates the amount of difference between two randomly drawn sample means that can be expected (from the normal curve ±1 *SD*) by chance alone.

Table 8.1   Research Design for Two Group Between Comparison.

Group	Number	Pretest	Treatment	Posttest
Control	N_1	Yes	No	Yes
Experimental	N_2	Yes	Yes	Yes

When t is larger than a given critical ratio in table A.3, we may conclude that (a) one or both of the samples were not randomly drawn, or (b) some factor has intervened to cause one or both of the groups to alter their mean value.

Now let us review the process for conducting experiments with two sample groups. Two groups of subjects are randomly selected from a population (or they may be selected by categories; e.g., age, gender). One group, the control group, is controlled very closely in all respects; the other group, the experimental group, is controlled in all respects except one. That one factor, the independent variable, is allowed to influence only the experimental group.

After these conditions are met, measurements of the dependent variable are taken on both groups. If the t ratio between the actual mean difference $(\overline{X}_1 - \overline{X}_2)$ and the expected mean difference (the standard error of the difference) is larger than the critical ratio for t (in table A.3) at a given p level, we conclude that the independent variable has had a significant effect, and we reject chance as a cause for the mean difference. The two groups are no longer representative of the same population from which they were drawn, because the experimental group has been changed by the independent variable.

Standard Error of the Difference

Equation 8.02, presented earlier, can be used to calculate the standard error of the mean (SE_M), or the amount that a single randomly drawn sample mean can be expected to deviate from a population mean by chance alone. A similar formula permits us to calculate the **standard error of the difference** (SE_D), the amount of difference between two randomly drawn sample means that may be attributed to chance alone.

When SE_M has been calculated for each of two sample means randomly drawn from the same population, we can use equation 8.05 to estimate the amount of the difference between the two means attributable to chance:

$$SE_D = \sqrt{(SE_{M1})^2 + (SE_{M2})^2}. \tag{8.05}$$

This formula (based on only two samples) estimates the size of the standard deviation of an infinitely large group of difference scores, each of which has been derived from randomly drawn pairs of sample means. Because it is a standard deviation, it may be interpreted in the same manner as any Z score on a normal curve.

Equation 8.06 presents the t test for independent samples:

$$t = \frac{\overline{X}_1 - \overline{X}_2}{SE_D}. \tag{8.06}$$

This t test is a ratio of the actual difference between two means (the numerator of t) and the difference that would be expected due to chance for an infinite set of sample means (the denominator of *t*). If the *t* ratio exceeds the critical ratio at a given *p* level, then chance, or H_0, is rejected as a probable cause for the difference between the sample means, and it is concluded that another factor or factors caused the difference (H_1 is accepted). We may then say that the mean difference is significant.

The statistics in the *t* test do not identify the causative factor. Careful controls and proper experimental design are needed to identify the factor. If the control group is closely monitored to assure that only chance operates on it, and if only one independent variable is permitted to influence the experimental group, then the independent variable may be identified as the causative factor.

To use table A.3 in appendix A for a two-group, independent *t* test, we must first calculate the degrees of freedom, *df*:

$$df = (N_1 - 1) + (N_2 - 1). \tag{8.07}$$

One degree of freedom is lost for each group. Then we compare the calculated *t* ratio with the critical ratios in table A.3 for the appropriate degrees of freedom. If the calculated *t* ratio exceeds the critical ratio for *t*, the mean difference is declared to be significant at the indicated *p* level, and H_0 is rejected.

An Example From Pedagogy

A pedagogy researcher wanted to know if a 10-minute verbal lesson in basketball shooting techniques would have any effect on the shooting ability of high school physical education students. The population for this study was all students at a given high school. The investigator randomly selected 100 students from a list of all students enrolled in the school. The subjects were then contacted by phone, and appointments were made with each one to meet in the gym dressed for activity.

The odd-numbered subjects (the control group) listened to a 10-minute verbal lesson on zone and one-on-one defense, warmed up for 5 minutes, and then made 20 free throws at the basket. The even-numbered subjects (experimental group) listened to a 10-minute verbal lesson on basketball shooting techniques, warmed up for 5 minutes, and then made 20 free throws at the basket.

The lesson on defenses for the control group equalized the time spent with the experimental group but was irrelevant to the shooting task. In both cases, the lesson was verbal, with no audiovisual equipment, and identical for each subject in the group. The following data on the number of successful shots out of 20 attempts were collected:

Control Group Defense lesson	Experimental Group Shooting technique lesson
$N_1 = 50$	$N_2 = 50$
$\overline{X}_1 = 9.5$	$\overline{X}_2 = 10.3$
$SD_1 = 2.7$	$SD_2 = 3.2$

Did the shooting lesson have an effect? The data presented previously show that there is a difference in favor of the experimental group, but it could be the result of random selection. If the experiment were to be repeated, but with the defense lesson given to both groups, a difference this large may still be obtained. A t test is needed to determine if the observed difference is random or significant.

To perform the t test, we need to calculate the standard error of the mean for each group and the standard error of the difference between the groups.

The standard error of the mean for the control group is

$$SE_{M1} = \frac{2.7}{\sqrt{50}} = .38.$$

And the standard error of the mean for the experimental group is

$$SE_{M2} = \frac{3.2}{\sqrt{50}} = .45.$$

Using equation 8.05, we calculate the standard error of the difference to be

$$SE_D = \sqrt{.38^2 + .45^2} = .59.$$

Remember, the standard error of the difference is the standard deviation of an infinite set of mean differences that form a normal curve. So its value ($SE_D = .59$) represents the value of the mean difference that would be expected 68% of the time between two randomly drawn samples influenced only by chance. This value is compared with the actual difference observed between the control and experimental means by a t test:

$$t = \frac{9.5 - 10.3}{.59} = -1.36.$$

The t value is negative because a larger value was subtracted from a smaller value in the numerator. The sign of the t ratio is not important in the interpretation of t, because it may be positive or negative depending on which group is listed first in the numerator. Only the absolute value of t is considered in determining significance.

The obtained t ratio (1.36) is compared to the critical ratios for a two-tailed test (one-tailed and two-tailed tests are explained later in this chapter) in table A.3 for $df = (50 - 1) + (50 - 1) = 98$. Table A.3 does not list a df value of 98, so we use the next lowest value, 60. For $df = 60$, the critical ratio at the most liberal p level, $p = .10$, is 1.671. Our obtained value of 1.36 does not reach this level. Because t is not significant at $p = .10$, we do not need to compare it at higher levels. We conclude that the differences could have happened by chance alone, so we accept the null hypothesis, H_0. The verbal shooting lesson had no significant effect.

An Example From Leisure Studies/Recreation

The *t* test can be applied to problems in many disciplines. A recreation director wanted to know if daily positive verbal reinforcement affected a person's ability to bowl. So the director conducted a 10-week experiment with two recreational bowling classes. In one class, the director praised the bowlers on their daily game scores, recorded the scores in a special book, and posted the scores on the wall.

In a second class, the director was neutral in relationships with the bowlers and made no comments, either positive or negative, on performance. The two groups were assumed to be equal in bowling ability at the start. To find out whether the praise had an effect on the average bowling scores at the end of the 10-week period, a *t* test was conducted on the following data:

Class 1 (praise)	Class 2 (neutral)
$N = 30$	$N = 30$
$\bar{X}_1 = 145$	$\bar{X}_2 = 135$
$SD_1 = 18.1$	$SD_2 = 14.9$

The standard error of the mean for the experimental group is

$$SE_{M1} = \frac{18.1}{\sqrt{30}} = 3.3.$$

For the control group,

$$SE_{M2} = \frac{14.9}{\sqrt{30}} = 2.7.$$

The standard error of the difference is

$$SE_D = \sqrt{(3.3)^2 + (2.7)^2} = 4.3.$$

This results in a *t* value of 2.33:

$$t = \frac{145 - 135}{4.3} = 2.33$$

$$df = (30 - 1) + (30 - 1) = 58$$

Table A.3 indicates that the *t* ratio of 2.33 reaches the $p < .05$ level for $df = 40$ (no value is listed for $df = 58$). In other words, the odds of finding a mean difference as large as 10 points by chance alone are less than 5 in 100. The director concluded that praise did make a difference with a level of confidence better than 95% (H_0 was rejected and H_1 was accepted). The results of a *t* test are often reported in tabular form as shown in table 8.2.

Table 8.2   Effects of Praise on Bowling Scores

Group	Mean	SD	SE_M	SE_D	t	p
Praise	145	18.1	3.3	4.3	2.33	< .05
Neutral	135	14.9	2.7			

The t Test With Unequal Values of N

The examples presented thus far have involved two groups with equal values of N. In practical research, this is almost never the case. Subjects often drop out of experiments (this phenomenon is interestingly referred to as subject mortality), and the two groups usually do not have equal numbers of subjects. The formula for standard error of the difference must be modified to account for the differences in N.

The standard error of the difference formula (equation 8.05) sums the two standard errors of the mean values based on the assumption that both values contribute equally to the standard error of the difference. This is true if $N_1 = N_2$. But if N_1 is twice as large as N_2, N_1 should contribute two-thirds of the total value of standard error of the difference. Yet equation 8.05 permits it to contribute only half. For this reason, the following alternate formula for standard error of the difference is used when values of N are unequal:

$$SE_D \sqrt{\left[\frac{(N_1 - 1)(SD_1)^2 + (N_2 - 1)(SD_2)^2}{N_1 + N_2 - 2}\right]\left[\frac{1}{N_1} + \frac{1}{N_2}\right]}. \quad (8.08)$$

This formula does not require the calculation of standard error of the mean for each group. Standard error of the difference is calculated directly from the standard deviations and N_1 and N_2. This saves one step in the process. Equation 8.08 produces the same answer as equation 8.05 when $N_1 = N_2$, so it could be used in every case.

When the values of N are large and only slightly unequal, the error introduced by using the simpler equation 8.05 to compute SE_D is probably not critical. But when the values of N_1 and N_2 are small, and approach a ratio of 2:1, the error introduced by equation 8.05 is considerable. If there is any doubt about which equation is appropriate, equation 8.08 should be used so that maximum confidence can be placed in the result.

An Example From Biomechanics

The following example applies equation 8.08 to the mean values obtained in a laboratory test comparing hip and low-back flexion of randomly selected males and females. The following measurements in centimeters were obtained using the sit-and-reach test:

Males	Females
$\bar{X}_1 = 22.5$	$\bar{X}_2 = 25.6$
$SD_1 = 2.5$	$SD_2 = 3.0$
$N_1 = 10$	$N_2 = 8$

Using equation 8.08, we calculate that $SE_D = 1.29$:

$$SE_D \sqrt{\left[\frac{(10-1)(2.5)^2 +(8-1)(3.0)^2}{10+8-2}\right]\left[\frac{1}{10}+\frac{1}{8}\right]}=1.29.$$

Once SE_D is known, *t* can be calculated:

$$t = \frac{22.5-25.6}{1.29} = -2.40.$$

The degrees of freedom are determined using equation 8.7. In this example $df = (10 - 1) + (8 - 1) = 16$. Table A.3 indicates that for $df = 16$, a *t* ratio of 2.40 is significant at $p < .05$. So H_0 is rejected, and H_1 is accepted. The researcher concludes with better than 95% confidence that females are more flexible than males in the hip and low-back joints as measured by the sit-and-reach test.

Repeated Measures Design (A Within Comparison)

The standard formulas for calculating *t* assume no correlation between the groups. Both groups must be randomly selected from the population and independent of each other. But when a researcher tests a group of subjects twice, such as in a pre-post comparison, the groups are no longer independent. **Dependent samples** assume that there is a relationship, or correlation, between the scores and that a person's score on the posttest is partially dependent on his or her pretest score.

This is always the case when the same subjects are measured twice. A group of subjects is given a pretest, subsequently treated in some way, and then given a posttest. The difference between the pretest and posttest means is computed to determine the effects of the treatment. This arrangement is often referred to as a repeated measures design or within comparison.

Because both sets of scores are made up of the same subjects, there is a relationship, or correlation, between the scores of each subject on the pre- and posttests. The differences between the pre- and posttest scores are usually smaller than they would be if we were testing two different groups of people. Two test scores of a single person are more likely to be similar than are the scores of two different people.

If there is a positive correlation between the two groups, a high pretest score is associated with a high posttest score. The same is true of low scores. Consequently, the difference between the two means will tend to be smaller with single group pre-post comparisons than with independent two-group comparisons. This may result in a false conclusion that there is no significant difference between the pretest and posttest means.

This same argument holds true for studies using matched pairs, pairs of subjects who are intentionally chosen because they have similar characteristics on the variable of interest. These matched pairs—sometimes called research twins—are then divided between two groups so that the means of the groups on the pretest are essentially equal. One group is treated, and the other group acts as control; then the posttest means are compared.

We expect the matched group means to have smaller differences than if the two groups were not matched on the pretest. In effect, we have forced the groups to be equal on the pretest so that posttest comparisons may be made with more clarity. The matched twins in each group may be considered to be the same person, and the correlation between them on the dependent variable can be calculated.

To accomplish this matching process, all the subjects are given a pretest and then ranked according to score. Using a technique sometimes referred to as the ABBA assignment procedure, the researcher places the first (highest scoring) subject in group A, the second and third subjects into group B, the fourth and fifth into group A, the sixth and seventh into group B, and so forth until all subjects have been assigned. The alternation of subjects into groups ensures that for each pair of subjects (1 and 2, 3 and 4, 5 and 6, etc.) one group does not always get the higher score of the pair.

This technique usually results in a correlation between the groups on the dependent variable and in smaller mean differences on the posttest. But because the two groups start with almost equal means on the pretest, it is easier to identify the independent variable as the cause of posttest differences.

If correlated samples are used (either the same subjects or matched pairs) and if no correction is made, the researcher may falsely conclude that there is no difference between the means, when in fact a real, or significant, difference does exist but is smaller than expected.

Correction for Correlated Samples

The correction for correlated samples is made in the formula for standard error of the difference (SE_D) because this value indicates the difference to be expected by chance alone. By adjusting the SE_D formula, we can regulate t to more correctly reflect any real difference that may exist in dependent samples.

The correction is made by factoring out, or subtracting, the effects of the correlation between the two samples in the formula for standard error of the difference:

$$SE_D = \sqrt{(SE_{M1})^2 + (SE_{M2})^2 - 2\,r\,(SE_{M1})\,(SE_{M2})}. \tag{8.09}$$

A positive value for r reduces the SE_D, increases t, and provides a greater chance of finding significance. When r is negative (which is very unlikely with matched pairs or repeated measures), t becomes smaller.

When r is zero, the term $2r\,(SE_{M1})\,(SE_{M2})$ becomes zero, and the formula reverts to its original form. Actually, equation 8.09 is the more generalized form of equation 8.05; however, $2r(SE_{M1})(SE_{M2})$ is usually not included as a component when the groups are independent, because r is assumed to be zero.

A researcher compared 30 high school students on vertical jump in inches before and after 3 weeks of leg strength development. The following example shows how an incorrect conclusion that no differences exist could be made on correlated samples of $N = 30$, if $r = .60$, and the correction in SE_D is not made. The degrees of freedom for a dependent t test are $N_{pairs} - 1$.

Suppose the following values resulted from the study:

Vertical jump [pretest]	Vertical jump [posttest]
$\bar{X}_1 = 15$	$\bar{X}_2 = 17.5$
$SE_{M1} = .9$	$SE_{M2} = 1.5$

If we do not correct for correlated samples, $SE_D = 1.75$:

$$SE_D = \sqrt{(.9)^2 + (1.5)^2} = 1.75.$$

And $t = -1.43$ (which is not significant at 29 df):

$$t = \frac{15 - 17.5}{1.75} = -1.43.$$

When we apply the correction,

$$SE_D = \sqrt{(.9)^2 + (1.5)^2 - 2\,(.6)\,(.9)\,(1.5)} = 1.20,$$

and

$$t = \frac{15 - 17.5}{1.20} = -2.08,\ p < .05.$$

At $df = 29$, the uncorrected t (−1.43) is not significant, but the corrected t (−2.08) is significant at $p < .05$. Because the subjects in both tests are the same people, a serious error would be made without the correction factor in the formula for SE_D. The problem of correcting for unequal N never arises with matched pairs or repeated measure designs, because the pairs are matched and the N are always equal.

Another method to compute t for correlated samples, which does not require the calculation of r, is called the direct difference method. This method is sometimes preferred because it is easier to calculate by hand. The formula for the direct difference method is

$$t = \frac{\Sigma D}{\sqrt{[N\Sigma D^2 - (\Sigma D)^2]/(N-1)}},$$
(8.10)

where D = the difference between the pre- and posttest scores for each subject and N = the number of pairs of scores. The results are identical with both methods.

An Example From Leisure Studies/Recreation

A graduate student in leisure studies wanted to know the short-term effect of a 4-day bicycle tour on the self-esteem of the participants. To measure self-esteem, the student administered the Cooper-Smith self-esteem survey to 45 bicycle riders immediately before and after the 4-day trip. The results are shown in table 8.3.

The correlation between the pre- and posttests is quite high (.92). Because of this high correlation, the standard error of the difference is very small (.37). This low error value permits the researcher to find significant differences between the two mean values ($df = 44$, $p < .01$). Based on this analysis, the graduate student rejected H_0 and concluded that the 4-day bicycle tour did have an effect on self-esteem. However, note that the mean difference was only 2.3 points (40.7 – 38.4). While this difference did not happen by chance, is it large enough to be meaningful? The next section will provide a method to answer this question.

Table 8.3   Effects of a 4-Day Bicycle Tour on Self-Esteem

Variable	Mean	SD	SE_M	SE_D	r	t
Pretest	38.4	6.1	.90	.37	.92	−6.35
Posttest	40.7	6.3	.94			

The Magnitude of the Difference (Size of Effect)

It is common to report the probability of error (*p* value) reached by the *t* ratio. Declaring *t* to be significant at *p* = .05 or some similar level only indicates the odds that the differences are real and that they did not occur by chance. This is often termed statistical significance. But we must also consider practical significance. If the values of *N* are large enough, if standard deviations are small enough, and especially if the design is repeated measures, statistically significant differences may be found between means that are quite close together in value. This small but statistically significant difference may not be large enough to be of much use in a practical application. How important is the size of the mean difference?

Thomas and Nelson (1996, p. 144) suggest the use of omega squared (ω^2) to determine the importance, or usefulness, of the mean difference. Omega squared is an estimate of the percentage of the total variance (the difference between the means) that can be explained by the influence of the independent variable (the treatment). For a *t* test, the formula for omega squared is

$$\omega^2 = \frac{t^2 - 1}{t^2 + N_1 + N_2 - 1}. \tag{8.11}$$

Applying equation 8.11 to the data from the earlier problem comparing male and female hip and low-back flexibility yields

$$\omega^2 = \frac{(-2.4)^2 - 1}{(-2.4)^2 + 10 + 8 - 1} = .21.$$

In this case, 21% of the differences between males and females in hip and low-back flexibility can be attributed to gender. The remaining 79% of the variance is due to individual differences among subjects, other unidentified factors, and errors of measurement.

How large must omega squared be before it is considered important? The answer to that question is not statistically based. Each investigator or consumer of the research must determine the importance of omega squared. In this example, it is meaningful to know that 21% of the variance can be explained. But gender clearly is not the only variable that affects hip and low-back flexibility. Other factors, unidentified in this study, are at work.

Another method of determining the importance of the mean difference is the **effect size** (*ES*), which may be estimated by the ratio of the mean difference over the standard deviation of the control group, or the pooled variance of the treatment groups if there is no control group:

$$ES = \frac{\overline{X}_1 - \overline{X}_2}{SD_{\text{Control}}}. \tag{8.12}$$

The control group is normally used as an estimate of the variance because it has not been contaminated by the treatment effect. In the example of the impact of praise on learning to bowl, the effect size is

$$ES = \frac{145 - 135}{14.9} = .67.$$

Jacob Cohen (1988, p. 21), as quoted in the work of Winer and others (1991, p. 122), proposes that *ES* values of .2 represent small differences; .5, moderate differences; and .8+, large differences. Winer's group (1991) also suggests that *ES* may be interpreted as a *Z* (standardized) score of mean differences. In the bowling example, an *ES* of .67 indicates that the effect of praise on learning to bowl was moderate.

The amount of improvement from the pretest to the posttest in repeated measures designs can be determined by assessing the percent of change. The following formula will determine the percent of change (improvement) between two repeated measures:

$$\text{Percent improvement} = \left(\frac{\overline{X}_2 - \overline{X}_1}{\overline{X}_1} \right) \times 100, \tag{8.13}$$

where $\overline{X}_1$ and $\overline{X}_2$ represent the pre- and posttest mean values.

In the example of the effects of participation in a 4-day bicycle tour on self-esteem (table 8.3), the pre-post improvement is small—only 6% ([40.7 – 38.4]/ 38.4 × 100 = 6%). Although the paired *t* value (–6.35) is significant at $p < .01$, the improvement analysis indicates that the change was minimal at best.

Omega squared, effect size, and percent improvement are important attributes to report when mean differences are studied. They provide additional information to the consumers of the research to assist them in determining the usefulness of the conclusions. These values may be more meaningful than the *p* value, especially if *p* just misses being significant (i.e., $p = .06$).

Type I and Type II Errors

Two kinds of errors can be made in accepting and rejecting hypotheses. A type I error is committed when the null hypothesis is true but is erroneously rejected (differences are found that in reality do not exist). A type II error is made when the null hypothesis is not true but is incorrectly accepted (the research fails to detect differences that really do exist).

Neophyte researchers are sometimes accused of making type I errors in their zeal to find significant differences. But failure to find a difference does not render the research worthless. It is just as important to know that differences do not exist as it is to know that differences do exist. The experienced and competent researcher is honestly seeking the truth, and a correct conclusion from a quality research project is valuable even if the research hypothesis is rejected.

Table 8.4 demonstrates the conditions under which type I and type II errors may be made. The dilemma facing the researcher is that one can never absolutely know which, if either, type of error is being made. The experimental design provides the means to determine the odds of each type of error, but complete assurance is never possible.

The researcher must decide which type of error is the most costly and then protect against it. If concluding that a difference exists when it does not (a type I error) is likely to risk human life or commit large amounts of resources to a false conclusion, then this is an expensive error and should be avoided. But if differences do exist, and the study fails to find them (a type II error), consumers of the research will never be able to take advantage of knowledge that may be helpful in solving problems.

Setting an appropriate *p* level to protect against either of the possible errors is critical. When using the null hypothesis, if we set a *p* level too low ($p = .10$ rather than $p = .05$, for example) and found the resulting *t* value to be significant at just barely $p = .10$, then we would reject the null hypothesis and conclude that a real difference exists. But in this case the odds are 10 out of 100 that the difference is not real. If there really is no difference between the means, and the resultant *t* just happens to be that 1-in-10 event that occurs by chance alone, we have committed a type I error.

It is also possible to err in the other direction. For example, if we use the null hypothesis with $p = .01$ and the *t* value is not high enough to reach the critical ratio in table A.3 (perhaps due to a small *N* or to measurement or other experimental errors), then we accept the null hypothesis. But if in reality the means do differ, we have committed a type II error.

Table 8.4   Type I and Type II Errors

	Reality	
	Null hypothesis is true	Null hypothesis is false
Accept null	No error; conclusion is correct.	Type II error; conclusion is incorrect.
Reject null	Type I error; conclusion is incorrect.	No error; conclusion is correct.

Key point !!

We can never know absolutely when we have made either of these kinds of errors. Statistical techniques only permit us to make probability statements about the truth. To reduce the probability of type I errors, use a more conservative p value ($p = .01$ instead of $p = .05$). To guard against type II errors, set a more liberal p value. Researchers must make a trade-off decision: Protecting against one type of error increases the probability of committing the other type.

The critical factor in this decision is the consequence of being wrong. The confidence level should be set to protect against the most costly error. We must ask, Which is worse: to accept the null hypothesis when it is really false or to reject it when it is really true?

look @ what's common to both types of error!

Possible Causes of Error	
Type I	Type II
1. Measurement error	1. Measurement error
2. Lack of random sample	2. Lack of sufficient power (N too small)
3. p value too liberal ($p = .10$)	3. p value too conservative ($p = .01$)
4. Investigator bias	4. Treatment effect not properly applied
5. Improper use of one-tailed test	

good slide & exam q fodder

Two- and One-Tailed Tests

Most research is done because the results of the experiment are not known beforehand. If the researcher can answer the research question through logical reasoning or a review of related literature, the experiment is not necessary. When review of all prior research does not yield an answer, the researcher proposes the null hypothesis, H_0, and conducts an experiment to test the hypothesis.

One way to state the null hypothesis is to predict that the difference between two population means is zero ($\mu_1 - \mu_2 = 0$) and that small differences in either direction (plus or minus) on the sample means are considered to be chance occurrences. The direction, or sign, of the difference is not important, because we do not know before we collect data which mean will be larger. We are simply looking for differences in either direction.

Two-Tailed Test

The null hypothesis is tested with a **two-tailed test.** If t does not reach the critical ratio for $p = .05$, then we know that the sample mean difference falls within the area of the normal curve that includes 95% of all possible differences, and we can accept H_0 with only a 5% chance of being wrong. When N values are large ($df > 120$), this corresponds to a Z score of ± 1.96 (see figure 8.1). In this case the 5% rejection

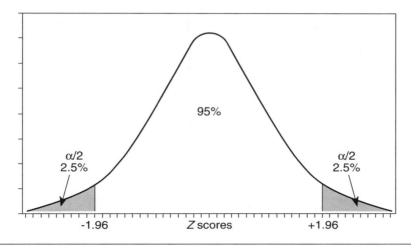

Figure 8.1   Distribution of alpha rejection area for a two-tailed test.

area (alpha) is divided between the two tails of the curve; each tail includes 2.5% of the area under the curve ($\alpha/2$).

Use the null hypothesis and a two-tailed test when prior research or logical reasoning does not clearly indicate that a significant difference between the mean values should be expected. Use the columns for a two-tailed test in table A.3 in appendix A to determine the *t* value needed to reject the null hypothesis at the predetermined *p* value.

One-Tailed Test

Sometimes the review of literature or logic strongly suggests that a difference does exist between two mean values. The researcher is confident that the direction of the difference is well established but is not sure of the size of the difference. In this case, the researcher may test the research hypothesis (H_1), but such a situation is rare. The evidence suggesting a mean difference must be strong to justify testing H_1. The opinion of the investigator alone is not sufficient.

The researcher predicts that two population means are not equal. By convention, the mean expected to be larger is designated as μ_1. Because the first mean is predicted to be greater than the second, the direction of the difference is established as positive.

The difference to be tested is always positive ($\mu_1 > \mu_2$), so we are interested in only the positive side of the normal curve. If an observational comparison of sample means shows $\bar{X}_2$ to be larger (even by the slightest amount) or equal to $\bar{X}_1$, the hypothesis that $\mu_1 > \mu_2$ is proven false and must be rejected.

When the direction of the difference is well established before data collection, use the research hypothesis and the one-tailed columns in table A.3 to determine the critical *t* ratio.

The **one-tailed test** places the full 5% of the alpha area representing error at one end of the curve (see figure 8.2). The *Z* score that represents this point (1.65) is lower than the *Z* score for a two-tailed test (1.96).

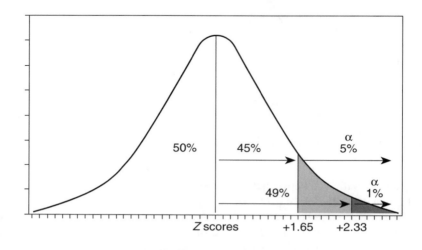

Figure 8.2   Distribution of alpha rejection area for a one-tailed test.

Table A.3 (we are assuming $df > 120$ for this discussion) shows that the *t* value is 1.65 for $p = .05$ in a one-tailed test (45% of the area under the curve). For $p = .01$ (49% of the area under the curve), the one-tailed *t* value is 2.33. Note that the *t* score does not need to be as large in a one-tailed test as in a two-tailed test (the $p = .01$ two-tailed *t* value is 2.58). Therefore, it is easier to find significant differences when a one-tailed test is used. For this reason, the one-tailed test is more powerful, or more likely to find significant differences, than the two-tailed test. But, it also presents a greater risk for a type I error. The next section discusses the power of a test and how it is calculated.

Determining Power and Sample Size

Power is the ability of a test to correctly reject a false null hypothesis. See Tran, Zung Vu, in *Measurement in Physical Education and Exercise Science*, vol. 1, no.1, 1997, p. 89, for an excellent discussion of the importance of calculating power. Because the critical *t* values are lower for a one-tailed test than for a two-tailed test (see table A.3), the one-tailed test is considered more powerful at a given *p* value. Also, a value of *p* = .05 is more powerful than *p* = .01 because the *t* value does not need to be as large to reach the critical ratio.

In figure 8.3, the range of the control group represents all the possible values for the mean of the population from which a random sample was taken. The most likely value is at the center of the curve, and the least likely values are at the extremes. The range of the experimental group represents all possible values of the population from which it was taken after treatment has been applied. The alpha point (Z_α) on the control curve is the point at which a null hypothesis is rejected for a given mean value in the experimental group.

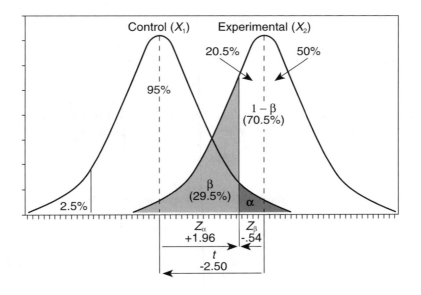

Figure 8.3   Calculation of power for an independent *t* test.

Any value for the mean of the experimental group that lies to the right of Z_α (1 − β area) will be judged significantly different from the mean of control at *p* < .05 (i.e., not taken from the same population). If H_0 is really true, this represents a type I error. Conversely, any value for the mean of the experimental group that lies to the left of Z_α (the beta area) will be judged to be not significantly different from the mean of control group. If H_0 is false, this represents a type II error.

It then follows that the area of 1 − β in the experimental group is the area of power, the area where a false null hypothesis will be correctly rejected. This area represents all of the possible values for the mean of the experimental population that fall beyond the Z_α level of the control population. Power is calculated by determining Z_β, converting it to a percentile using table A.1 in appendix A, and adding this percent of area to the 50% of the curve to the right of the experimental mean.

In figure 8.3, $1 - \beta$ represents 70.5% of all the possible mean values for the experimental group. So there is a 70.5% chance that a false null hypothesis will be rejected; power = 70.5%. Let's consider how power is calculated.

As figure 8.3 shows, power is dependent on four factors:

1. The Z_α level set by the researcher (the level set to protect against type I errors; $p = .05$, $p = .01$, etc.). It is represented by a Z_α score from the normal distribution table $[(A.1), (Z_\alpha (.10) = 1.65, Z_\alpha (.05) = 1.96, Z_\alpha (.01) = 2.58)]$.
2. The difference, Δ ($\Delta = \bar{X}_1 - \bar{X}_2$), between the two mean values being compared (where $\bar{X}_1$ is the mean of the control group and $\bar{X}_2$ is the mean of the experimental group).
3. The standard deviations of the two groups (SD), which determine the spread of the curves.
4. The sample size, N, of each of the two groups.

Only N and Z_α are under the control of the researcher, and Z_α usually cannot be radically manipulated because of the need to protect against type I errors. Therefore, the researcher can control power primarily by manipulating the size of N.

Calculating Power

The following process is used to calculate power—that is, to determine the $1 - \beta$ area in figure 8.3.

The researcher sets Z_α, the p value set to reject the null hypothesis. The values for the means, standard deviations, N for each group, the standard error of the difference, and t are calculated. Figure 8.3 demonstrates that t is the sum of Z_α and Z_β. To determine the power area of the experimental curve, we must find the value of Z_β, which is the percent of the area on the experimental curve between $\bar{X}_2$ and Z_β.

In figure 8.3, Z_α is a positive value; it proceeds to the right of the control mean ($\bar{X}_1$). The t value is negative because $\bar{X}_1 < \bar{X}_2$. The Z_β value is also negative; it proceeds to the left of the experimental mean X_2. If the analysis were made with $\bar{X}_1 > \bar{X}_2$ then t and Z_β would be positive values, and Z_α would be negative. In order for the following formulas to be applied toward either tail of the curve, the values of t, Z_α, and Z_β will all be considered absolute. Then t is equal to the sum of Z_α and Z_β:

$$t = Z_\alpha + Z_\beta. \tag{8.14}$$

Conversely,

$$Z_\beta = t - Z_\alpha. \tag{8.15}$$

Let us assume the following data apply to figure 8.3:

$\bar{X}_1 = 30$, $\bar{X}_2 = 32.5$ ($\Delta = 2.5$)

$SD = 5$ for each group

$N = 50$ for each group

$SE_M = .71$ for each group, ($5 / \sqrt{50} = .71$)

$SE_D = 1.00$, ($\sqrt{.71^2 + .71^2} = 1.00$)

$t = 2.50$, ($2.5 / 1.0 = 2.5$)

$Z_\alpha = 1.96$ ($p = .05$)

Equation 8.15 may now be used to determine Z_β:

$$Z_\beta = 2.50 - 1.96 = .54.$$

We convert the Z_β value of .54 to a percentile from table A.1 ($Z_\beta = .54 = 20.5\%$ of the area under the normal curve) and compute the area of $1 - \beta$ ($20.5\% + 50\% = 70.5\%$). Therefore 70.5% of all possible values of the experimental population mean lie to the right of Z_α. In other words, there is a 70.5% chance of rejecting the null hypothesis if the values given in the data section above are true; power = 70.5%.

Calculating Sample Size

The only factor in these equations that is easily manipulated by the researcher is N. We could increase our power by increasing N, but how large does N need to be to produce a given power level?

Equation 8.14 may be solved for N as follows. The formula for t is

$$t = Z_\alpha + Z_\beta.$$

Because $t = \Delta / SE_D$, we may substitute Δ / SE_D for t.

$$\frac{\Delta}{SE_D} = Z_\alpha + Z_\beta.$$

Recall that for $N_1 = N_2$ (equation 8.05),

$$SE_D = \sqrt{(SE_{M1})^2 + (SE_{M2})^2}$$

and (equation 8.02)

$$SE_M = \frac{SD}{\sqrt{N}}$$

When $N_1 = N_2$ and $SD_1 = SD_2$:

$$SE_D = \sqrt{\left(\frac{SD_1}{\sqrt{N_1}}\right)^2 + \left(\frac{SD_2}{\sqrt{N_2}}\right)^2} = \sqrt{\frac{2SD^2}{N}}.$$

Substituting this value for SE_D, we obtain the following:

$$\frac{\Delta}{\sqrt{\frac{2SD^2}{N}}} = Z_\alpha + Z_\beta$$

$$\Delta = \left(\sqrt{\frac{2SD^2}{N}}\right)(Z_\alpha + Z_\beta)$$

$$\Delta^2 = \left(\frac{2SD^2}{N}\right)(Z_\alpha + Z_\beta)^2$$

$$N\Delta^2 = 2SD^2 (Z_\alpha + Z_\beta)^2$$

Therefore,

$$N = \frac{2SD^2 (Z_\alpha + Z_\beta)^2}{\Delta^2}. \tag{8.16}$$

With equation 8.16, we can determine the N needed for a given power level if we know the other values. Suppose we want power = .80 at $p = .05$. Then the area of Z_β must be 30% and $1 - \beta$ is 80%. We look up 30% in table A.1 to find that $Z_\beta = .84$. If $\Delta = 5$, and for each group $SD = 6$, and we set Z_α at 1.96, then

$$N = \frac{2(6)^2 (1.96 + .84)^2}{5^2} = 22.6.$$

We conclude that to achieve 80% power under these conditions, we must use a sample size of approximately 23 in each group.

The calculation of power is a major factor in experimental design. It is important to know what the odds are that real differences between group means may be detected before we conduct expensive and time-consuming research. Research performed with insufficient power (i.e., N is too small) may result in a type II error (failure to reject a false null hypothesis) or may waste valuable resources on a study that has little chance of rejecting the null.

In a power calculation, the values for the means and standard deviations are not usually known beforehand. To calculate power before the data are collected, these values must be estimated from pilot data or from prior research on similar subjects.

The previous power calculation example is applicable only to a *t* test of independent means, with equal values of both *N* and *SD*. This is the most simple application of the concept of power. Similar calculations may be made for unequal values of *N* or for dependent tests.

A software program titled *PC Size* is described by Dallal (1986). It can be used to perform power calculations for simple research designs. Additional discussions of power may be found in Kachigan (1986, p. 185), and Thomas and Nelson (1996, p. 110).

The *t* Test for Proportions

The techniques for estimating error in sample means and determining the significance of the difference between two means may be modified to apply to proportions. The following example illustrates this procedure.

An Example From Administration

A teacher surveyed 150 girls in a large school to determine their favorite subject and found that 60% chose physical education. The principal doubted these findings and claimed that the true population value of those favoring physical education couldn't be more than 50%. He challenged the 60% figure and asked the teacher to prove it.

If we assume that the teacher's survey was properly conducted and that subjects were randomly drawn from the population, what are the odds that another random survey from the same population could result in a value of 50%? To answer such a question, we need to know the error that can be expected in a proportion. This error can be estimated with the following formula for the standard error of a proportion,

$$SE_p = \sqrt{\frac{pq}{N}}, \qquad (8.17)$$

where *p* is the obtained proportion, $q = 1 - p$, and N = sample size. Applying this formula to the problem at hand yields

$$SE_p = \sqrt{\frac{.60\,(1 - .60)}{150}} = .04.$$

The standard error of a proportion (SE_p) may be interpreted as a Z score. Therefore, the odds that the true proportion lies between .56 and .64 (.60 ± .04) are 68 to 32. Multiplying SE_p by 1.65 ($p = .10$), 1.96 ($p = .05$), or 2.58 ($p = .01$) will produce the limits of the population at a given level of confidence. For example, at $p = .05$, .60 ± (1.96).04 = .60 ± .078 indicating that the odds that the true population mean lies between .522 and .678 are 95 to 5.

Based on this analysis, the principal agreed that the true proportion for the population was probably not 50%.

This concept may be applied to a t test between two proportions (Bruning & Kintz, 1977, p. 222). The formula for a t test between proportions, t_p, is

$$t_p = \frac{P_1 - P_2}{\sqrt{\dfrac{p(1-p)}{N_1} + \dfrac{p(1-p)}{N_2}}}, \tag{8.18}$$

where P_1 and P_2 are the proportions to be compared, and p under the radical is

$$p = \frac{N_1 P_1 + N_2 P_2}{N_1 + N_2}. \tag{8.19}$$

The t test for proportions should not be used when either p or q times N is less than 5. Under these conditions, use nonparametric statistics (see Witte, 1985, p. 155).

An Example Comparing Two Proportions

If 60% of 150 girls and 70% of 125 boys chose physical education as their favorite subject, is there a significant difference between the girls and the boys at $p = .05$? To answer this question, we calculate

$$p = \frac{(150)(.60) + (125)(.70)}{150 + 125} = .65$$

and

$$t_p = \frac{.60 - .70}{\sqrt{\dfrac{.65(1-.65)}{150} + \dfrac{.65(1-.65)}{125}}} = 1.74.$$

Then we look in table A.3 to interpret t_p for $df = N_1 + N_2 - 2$. The t value of 1.74 is not large enough to reject chance at $p = .05$ because it does not reach the critical ratio of 1.96 for $df = 273$. There is no significant difference between the proportion chosen by the girls and the proportion chosen by the boys.

Summary

It is very unlikely that means of two random samples from the same population will be identical. Differences will almost always be observed. This is not unexpected; people do not always perform exactly the same, and even if they did, the measurement of their performance is not perfect. Because of these random errors, we always expect mean values to differ. The question is, how much can they differ before we suspect that the difference is caused by something other than chance?

As we discussed earlier in this chapter, the purpose of a *t* test is to determine whether the differences between two mean values are large enough to reject chance as a probable cause. Using the concepts of the normal curve, we can determine the amount of difference between any two means that can be attributed to chance alone. If the observed difference is larger than this estimated difference, then we reject chance as a cause and look for another reason to explain the mean difference.

The *t* test is the technique by which we perform this analysis. The *t* value is simply the ratio of the observed difference (the numerator) to the expected difference (the denominator). If the observed difference is larger than the expected difference, the *t* ratio can be compared to the critical ratios in table A.3 to determine the probability that the observed difference occurred by chance. The table values, or critical ratios, are the values of *t* for selected sample sizes, or degrees of freedom, that would be expected to occur by pure chance. When our obtained *t* exceeds these values, we reject the null hypothesis (H_0) and declare the differences to be significant (i.e., not due to chance).

The *t* test is useful for conducting experimental research. If we want to know the effect of some treatment, we compare a group that has had the treatment with one that has not. If the treatment is ineffective, we expect only chance differences between the groups. If the treatment is effective, the differences will exceed the expected difference. The *t* test may be modified to make comparisons between observed proportions.

Following is a list of essential steps that need to be conducted to properly determine the significance of the difference between two population means using randomly selected groups.

1. Define the population of interest.
2. State the problem.
3. Review the literature. Determine if the problem is solved by prior research.
4. If the problem is not solved, state a hypothesis (H_0 = null, H_1 = directional).
5. Select a level of confidence (consider consequences of error).
6. Select a power level and determine appropriate sample size.
7. Randomly select two samples from the population. Treat one sample with the independent variable. Provide appropriate controls for the other sample.
8. Compute mean, standard deviation ($N - 1$), and standard error of the mean for each sample.

9. Compute standard error of the difference and t.
10. Determine degrees of freedom.
11. Compare obtained t to critical values in a table of t or read p value from computer.
12. Make a conclusion by accepting or rejecting the hypothesis at the given level of confidence.
13. Determine the practical importance of the conclusion by calculating the size of the effect.

Problems to Solve

1. Calculate the standard error of the difference for each of the following sets of data.

	$\bar{X}_1$	$\bar{X}_2$	N_1	N_2	SD_1	SD_2
A.	172	175	50	50	20	18
B.	9.7	7.0	10	15	2.5	3.1

2. What are the t values for problems 1A and 1B? Is either significant? At what level of confidence?

3. Give a verbal definition of the meaning of the standard error of the difference.

4. In a study of absolute errors in active versus passive arm positioning, an investigator collected data in centimeters on 20 college-age subjects (from the motor learning laboratory, California State University Northridge, courtesy of Tami Abourezk). Is there a significant difference in the errors made by the active group ($N_1 = 10$) versus the passive group ($N_2 = 10$) on arm positioning?

Subject	Active	Subject	Passive
1	2.65	11	3.30
2	2.42	12	2.00
3	3.30	13	0.09
4	0.19	14	0.04
5	1.25	15	4.56
6	2.00	16	3.33
7	3.34	17	1.02
8	4.08	18	0.89
9	0.70	19	2.78
10	2.89	20	1.65

5. A graduate student in biomechanics was interested in stride length of cross-country skiers. Stride length is an important factor in the development of speed for racing. Data were collected on 20 athletes from the cross-country ski team (the experimental group). The data were compared to those of a second group of 17 students (the control group) who were not varsity athletes but did participate in recreational skiing. The researcher assumed that no significant differences would be found. Is there a significant difference between the athletes and the nonathletes in stride length? What is the level of confidence? What is ω^2, and what is the effect size? (Adapted from Duoos, 1984, data fabricated).

Athletes	Nonathletes
$\bar{X}_1 = .90$ meters	$\bar{X}_2 = .70$ meters
$N_1 = 20$	$N_2 = 17$
$SD_1 = .17$	$SD_2 = .23$

6. Find the *t* value for the following data on dominant versus nondominant grip strength. Twenty subjects were measured twice, once with the dominant hand and once again with the nondominant hand. Which hypothesis, research or null, might be appropriate in this study? Is the difference significant, and if so, at what level of confidence?

Dominant		Nondominant
$\bar{X}_1 = 40$ pounds		$\bar{X}_2 = 35$ pounds
$SD_1 = 12.7$	$r = .83$	$SD_2 = 14.2$

7. A researcher in exercise physiology wanted to know if body composition differs among prepubescent males and females. To test the null hypothesis, she measured skinfolds in millimeters on 5 males and 6 females with the following results.

Males	Females
21	22
25	19
19	18
17	24
18	21
	23

A. Is the difference in the means significant? If so, at what level of confidence?

B. Do you accept or reject the null hypothesis?

8. A sport psychologist wondered if motivation could improve aerobic capacity. To answer the question, he measured 10 subjects on max VO_2 (ml/kg/min) on a treadmill. Students were instructed verbally to "do your very best." One week later, the subjects were measured again but this time they were told they would receive $100 if their max VO_2 was higher than the first time. Following are the data.

First	Second
45	54
33	50
59	58
32	38
30	42
27	35
29	38
59	66
44	48
40	49

A. Did the money cause them to significantly raise their max VO_2 values?
B. If so, what is the probability of error in your conclusion? Should you accept or reject the null hypothesis?

9. An aerobic dance teacher wanted to know if two workouts per week of 30 minutes each were enough to increase VO_2max in sedentary middle-aged women. The teacher proposed to compare VO_2max in some women who had been in the aerobic dance classes for 6 months to another group of sedentary women. Based on related literature, the teacher estimated that VO_2max would be about 7 milliliters per kilogram per minute higher in the aerobic dancers, with a standard deviation of 4.5. If this is a fair estimate of the data to be expected, how many women must she test in each group to produce a power coefficient of .90 at $p = .05$?

See appendix C for answers to problems.

Key Words in This Chapter

Critical ratio	Dependent sample
Actual mean difference	Effect size
Expected mean difference	Two-tailed test
Independent sample	One-tailed test
Standard error of the difference	Power

Chapter 9

Simple Analysis of Variance: Comparing Means Among Three or More Sets of Data

An exercise physiologist interested in the relative effects of diet and exercise on body composition conducted a study with four groups: A control group received no treatment, a diet group followed a diet but no exercise program, an exercise group followed an exercise program but no diet, and a combined group followed both a diet and an exercise program. Percent body fat of the subjects in all four groups was measured at the beginning of the study (the pretest) and again after 6 weeks (the posttest). If the researcher wanted to analyze the data using the *t* test, six tests would be required to make all possible group comparisons on the posttest. Is there a better method?

Analysis of variance (ANOVA) is a parametric statistical technique used to determine whether significant differences exist among means of three or more sets of data. In a t test, the differences between the means of two groups are compared with the difference expected by chance alone. Analysis of variance compares the variability among three or more group means, the **between-group variability,** with the variability of scores within the groups, the **within-group variability.** This produces a ratio value called F (F = average variance between groups divided by average variance within groups). The symbol for analysis of variance (F) is named after the English mathematician Ronald Aylmer Fisher (1890–1962), who first described it (Kotz & Johnson, 1982, Vol. 3, p. 103).

If the between-group variability exceeds the within-group variability by more than would be expected by chance alone, it may be concluded that at least one of the group means differs significantly from another group mean. The null hypothesis (H_0) for an F test is designated as

$$\mu_1 = \mu_2 = \mu_3 = \ldots = \mu_k.$$

The null hypothesis assumes that the means of two or more populations are equal. Therefore, means of samples randomly drawn from the same populations should not differ by more than chance. When at least one sample mean is significantly different from any other, F is significant and we reject the null hypothesis.

Like the t test, the theoretical concepts of analysis of variance are based on random samples drawn from a population and the characteristics of the normal curve. If a large number of samples are randomly drawn from a population, the variance among the scores of all subjects in all groups is the best estimate of the variance of the population. When the variance among all the scores in a data set is known, it may be used to determine the probability that a deviant score is not randomly drawn from the same population. This argument may be expanded to infer that if randomly drawn scores are randomly divided into subgroups, the variance among the subgroup means may be expected to be of the same relative magnitude as the variance among all of the individual scores that comprise the groups.

With untreated data, when only random factors are functioning between the group means and within the scores of the groups, the between-group and within-group variances should be approximately equal. The F ratio, the ratio of the average between-group variance divided by the average within-group variance, is expected to be about 1.00. When the value of the F ratio exceeds 1.00 by more than would be expected by chance alone, the variance between the means is judged to be significant (i.e., the difference was caused by a factor other than chance).

The F ratio is analogous to the t ratio in that both compare the actual, or observed, mean differences with differences expected by chance. When this ratio exceeds the limits of chance at a given level of confidence, chance as a cause of the differences is rejected. In the F test, the actual differences are the variances between the group means, and the expected differences are the variances within the individual scores that make up the groups. The t test is actually a special case of analysis of variance with two groups. Because t uses standard deviations and ANOVA uses variance (V) to evaluate mean differences, and $SD^2 = V$, when there are only two groups in ANOVA, $t^2 = F$.

Analysis of variance is one of the most commonly used statistical techniques in research. But students often ask, Why is this new technique needed when a t test could be used between each of the groups? For example, in a four-group study, why not conduct six t tests—between groups A and B, A and C, A and D, B and C, B and D, and C and D? There are three reasons why multiple t tests are not appropriate.

1. **There is a greater probability of making a type I error (rejecting the null hypothesis when it is really true) when one conducts multiple t tests on samples taken from the same population.** When a single t test is performed, the findings are compared to the probability of chance. If a confidence level of 95% is set, we are willing to refute chance if the odds are 19 to 1 against it. When multiple t tests are conducted on samples randomly drawn from the same population, the odds of finding the one deviant conclusion that is expected by chance alone increase.

When 20 t tests are performed at $p = .05$ on completely random data, it is expected that one of the tests will be found significant by chance alone (1 to 19 odds). Therefore if we perform 20 t tests on treated data, and 1 of the 20 tests produces a significant difference, we cannot tell if it represents a true difference due to treatment or if it represents the one deviant score out of 20 that is expected by chance alone. If it is due to chance but falsely declared to be significant, a type I error has been made. This dilemma is sometimes referred to as the **familywise error rate** (the error rate when making a family of comparisons).

Keppel (1991, p. 164) reports that the relationship between the single comparison error rate (α) and the familywise error rate (FW_α) is

$$FW_\alpha = 1 - (1 - \alpha)^C, \qquad (9.01)$$

where C is the number of comparisons to be made. If we conduct six t tests, comparing all possible combinations of four groups (A, B, C, and D), at $\alpha = .05$, then

$$FW_\alpha = 1 - (1 - .05)^6 = .26.$$

In this example, conducting multiple t tests raises the probability of a type I error from .05 to .26. Keppel suggests that FW_α may be roughly estimated by the product of C and α. In this example, $FW_\alpha \cong 6 \times .05 = .30$. This method will always overestimate FW_α but is fairly close for small values of C and α. ANOVA eliminates this problem by making all possible comparisons among the means in a single test.

There may be occasions when multiple tests on samples from the same population are required. When this is the case, a commonly used modification of the alpha level called a **Bonferroni adjustment** is recommended. To perform the adjustment, divide the single-test alpha level by the number of tests to be performed. If five tests are to be made at $p = .05$, the adjusted alpha level to reject H_0 would be .01 (.05/5 = .01).

2. **The t test does not make use of all available information about the population from which the samples were drawn.** The t test is built on the assumption that only two groups have been randomly selected from the population. In the t test, the estimate of the standard error of the difference between means is based on data from two groups only. When three or more groups have been selected, information about the population from three or more samples is available that should be used in the analysis, yet t considers only two samples at a time. Analysis of variance employs all of the available information from all groups simultaneously.

3. **Multiple t tests require more time and effort than a simple ANOVA.** It is easier, especially with a computer, to conduct one F test than to conduct multiple t tests.

Because of these reasons, analysis of variance is employed in place of multiple t tests when three or more groups of data are involved. Analysis of variance can determine if a significant difference exists among any of the groups represented in the experiment, but ANOVA does not identify the group or groups that differ. A significant F value only indicates that at least one significant difference exists somewhere among the many possible combinations of groups. When a significant F is found, additional post hoc (after the fact) tests must be performed to identify the group or groups that differ. If F is not significant, no further analysis is needed because we know that there are no significant differences among any of the groups.

Assumptions in ANOVA

The F test is based on the following assumptions:

- The population from which the samples are drawn is normally distributed. Violation of this assumption has little effect on the F value among the samples (Keppel, 1991, p. 97). The F test produces valid results even when the population is not normally distributed. For this reason it is considered to be robust.
- The variability of the samples in the experiment is equal or nearly so (homogeneity of variance). As with the assumption of normality, violation of this assumption does not radically change the F value. However, as a general rule, the largest group variance should not be more than two times the smallest group variance.

- The scores in all the groups are independent; that is, the scores in each group are not dependent on, not correlated with, or not taken from the same subjects as the scores in any other group. The samples have been randomly selected from the population and randomly assigned to conditions. If there is a known relationship among the scores of subjects in the several groups, use repeated measures analysis of variance (see chapter 10).
- The data are based on a parametric scale, either interval or ratio. (For nonparametric data analysis, see chapter 13.)

The F test, like the t test, is considered robust. It provides dependable answers even when there are violations of the assumptions. Violations are more critical when sample sizes are small or Ns are not equal. If violations are committed that cannot be controlled and that the researcher thinks may increase the possibility of a type I error, a more conservative p value should be used to compensate for the violations (i.e., use $p = .01$ rather than $p = .05$).

Sources of Variance

The computation of F is simple in theory and does not involve difficult mathematics. When several groups of data are compared, each group has a mean (the group mean) and the entire data set, all groups combined, has a mean (the grand mean).

The grand mean is computed by summing the scores from all groups and dividing by the total number of subjects in all groups combined (N). Each of the group means may differ from the grand mean. The variance, or deviation, of the group means from the grand mean is called the **between-group variance.**

In addition to variance between the means, each individual score deviates from the mean of its group by a certain amount. This source of variance is called **within-group variance.** These two sources of variance, between-group and within-group, are the basic components used to compute the analysis of variance, F.

A third source of variance, the **total variance,** may be computed by determining the deviation of each score in each group from the grand mean. Total variance is equal to the sum of between-group variance and within-group variance:

$$\text{Variance}_{\text{total}} = \text{Variance}_{\text{between}} + \text{Variance}_{\text{within}}.$$

Total variance and between-group variance are relatively easy to calculate. But calculating within-group variance, although not mathematically difficult, is tedious. Because total variance is the sum of between- and within-group variance, within-group variance may be determined by subtracting between-group variance from total variance. Total variance is not essential for determining F, but it is useful in calculating within-group variance.

Sum of Squares and Mean Square

The **sum of squares** (*SS*) is the sum of the squares of the deviations of each score from a mean:

$$SS = \Sigma (X - \overline{X})^2 .$$

It is computed by subtracting each score from the mean, squaring the deviation scores, and adding them up. The sum of squares *within* any group can be computed from the individual scores and the group mean:

$$SS_W = \Sigma (X - \overline{X}_{group})^2 .$$

The sum of squares between groups can be computed by subtracting the grand mean from each group mean:

$$SS_B = \Sigma (\overline{X}_{group} - \overline{X}_{grand})^2 .$$

The sum of squares for the total can be computed by subtracting each individual score and the grand mean:

$$SS_T = \Sigma (X - \overline{X}_{grand})^2 .$$

The total sum of squares is always equal to the between-group sum of squares plus the within-group sum of squares:

$$SS_T = SS_B + SS_W . \tag{9.02}$$

In ANOVA, the size of the sum of squares between groups is compared to the size of the sum of squares within groups. The size of the sum of squares is dependent on (a) the number of scores summed and (b) the size of the squared deviations from the mean, or the variance. To account for the differences in the number of scores that make up the between-group (number of groups) and within-group (number of individual scores) sums of squares, the **mean square** (*MS*) is computed by dividing each *SS* by the appropriate degrees of freedom (*MS* = *SS/df*). This process makes the mean, or average, variabilities for SS_B and SS_W comparable.

Degrees of freedom within are determined by subtracting the number of groups (*k*) from the total number of subjects in all groups (*N*):

$$df_W = N - k . \tag{9.03}$$

Degrees of freedom between are determined by the number of groups (*k*) minus 1:

$$df_B = k - 1 . \tag{9.04}$$

To determine mean square within, divide SS_W by the degrees of freedom within:

$$MS_W = SS_W / df_W. \tag{9.05}$$

To determine mean square between, divide SS_B by the degrees of freedom between:

$$MS_B = SS_B / df_B. \tag{9.06}$$

These mean square values are now directly comparable and can be used to calculate F:

$$F = \frac{MS_B}{MS_W}. \tag{9.07}$$

The mean square within (MS_W) is the denominator of the F test. Like SE_D in the t test, it represents the amount of variance that can be expected due to chance occurrences alone. The F ratio compares MS_W to the differences between the means represented by mean square between (MS_B).

In ANOVA, MS_W is often referred to as mean square error and may be denoted MS_E. The term *error* does not mean a mistake; it means the variance that can occur by chance alone, or the variance that is unaccounted for by the effects of treatment. From this point on, we will use the representation MS_E, but it is synonymous with MS_W.

When MS_E is large, it tends to mask small treatment effects. When MS_E is small, it is easier to identify treatment effects. *The intent of all research designs is to keep MS_E as small as possible.* The value of MS_E depends on the variability in the data and on the number of scores (N) that make up the data. Of these two factors, only N can be controlled by the researcher. Studies with large values of N are more powerful in detecting mean differences.

Calculating F

We can calculate F using the formulas presented thus far. The concepts for this calculation of F are easy to understand because they are based on the definition of terms we have already introduced. But, like the definition formulas for standard deviation and for Pearson's correlation, the definition method for ANOVA is tedious to use when raw scores contain decimal values and N is large. A second computational method, the raw score method, is easier to use. The definition method is almost never used in practical applications but is presented here to explain the concepts that underlie ANOVA. The raw score method should be used to solve actual problems. Both methods, when correctly applied, produce the same answers.

The Definition Method

Table 9.1 presents hypothetical strength measurements on five groups (X_1 to X_5) of 7 subjects each. The numerical values do not represent any particular unit of measure. They are purposely small and discrete so that the calculations will be easy. The subjects were randomly selected from a population and randomly assigned to groups. Each subject completed 6 weeks of strength training. In this table, n is the number of subjects in each group, while N is the total number of subjects or scores in all groups combined. We shall test the null hypothesis (H_0) that none of the treatments had an effect. H_0 predicts no significant differences among any of the group means.

Groups 1, 2, 3, and 4 participated in different strength training programs; group 5 was the control group. We want to know whether any differences exist between the mean scores of the groups after the differential training. In this case, using ANOVA to solve the problem takes the place of 10 separate t tests.

The steps for calculating F by the definition method are as follows:

1. Calculate the sum of each group (ΣX_1, ΣX_2, ΣX_3, . . .), the mean of each group ($\overline{X}_1$, $\overline{X}_2$, $\overline{X}_3$, . . .), the grand sum (ΣX_T), and the grand mean (M_G). These values can be found at the bottom of table 9.1.
2. Calculate the within-group sum of squares (SS_W) for each group.
   - Determine the deviation scores of each raw score from its group, mean ($d = X - \overline{X}_{group}$), as shown in table 9.2.

Table 9.1   Data for Simple ANOVA

	X_1	X_2	X_3	X_4	X_5 (control)
	4	5	5	8	5
	5	7	4	4	4
	6	9	6	6	3
	7	8	5	8	4
	4	9	5	5	6
	6	7	6	6	4
	5	10	4	7	5
ΣX	37	55	35	44	31
n	7	7	7	7	7
$\overline{X}$	5.29	7.86	5.00	6.29	4.43

$\Sigma X_T = 37 + 55 + 35 + 44 + 31 = 202$
$N = 7 + 7 + 7 + 7 + 7 = 35$
$M_G = 202/35 = 5.77$

Table 9.2   Calculation of Within-Group Deviations

d_1	d_2	d_3	d_4	d_5
−1.29	−2.86	0.00	1.71	.57
−.29	−.86	−1.00	−2.29	−.43
.71	1.14	1.00	−.29	−1.43
1.71	.14	0.00	1.71	−.43
−1.29	1.14	0.00	−1.29	1.57
.71	−.86	1.00	−.29	−.43
−.29	2.14	−1.00	.71	.57

Note: These values are derived by subtracting the group mean from table 9.1 from the individual score (for the first score (X) in group X_1: $4.00 - 5.29 = -1.29$).

- Square each deviation and find the sum of d^2 for each group, as shown in table 9.3.
- Sum the squared deviations from each group. This value is the sum of squares within:

$$SS_W = \Sigma\Sigma(X - \overline{X}_{group})^2 = 7.40 + 16.86 + 4.00 + 13.40 + 5.71 = 47.37.$$

3. Calculate the between-group sum of squares (SS_B).
   - Find the deviation (d) of each group mean from the grand mean, square each deviation, and sum these deviations (Σd^2_B) (table 9.4).
   - Because there are n times more values contributing to SS_W than to SS_B, multiply Σd^2_B by n (7) to make SS_B directly comparable with SS_W:

$$SS_B = 7.26 \times 7 = 50.82.$$

4. Determine the degrees of freedom between and within:

$$df_B = 5 - 1 = 4, \quad df_E = 35 - 5 = 30.$$

5. Determine the mean square between (MS_B) and the mean square error (MS_E) by dividing the SS values by the appropriate df:

$$MS_B = 50.82 / 4 = 12.71, \quad MS_E = 47.37 / 30 = 1.58.$$

6. Determine the ratio (F) between MS_B and MS_E:

$$F = 12.71 / 1.58 = 8.04.$$

Table 9.3   Squaring and Summing Within-Group Deviations

$(d_1)^2$	$(d_2)^2$	$(d_3)^2$	$(d_4)^2$	$(d_5)^2$
1.66	8.18	0.00	2.92	.33
.08	.74	1.00	5.24	.18
.50	1.30	1.00	.08	2.05
2.92	.02	0.00	2.92	.18
1.66	1.30	0.00	1.66	2.46
.50	.74	1.00	.08	.18
.08	4.58	1.00	.50	.33
$\Sigma(d_1)^2 = 7.40$	$\Sigma(d_2)^2 = 16.86$	$\Sigma(d_3)^2 = 4.00$	$\Sigma(d_4)^2 = 13.40$	$\Sigma(d_5)^2 = 5.71$

Table 9.4   Calculation of Between-Group Deviations

	$\bar{X}$	M_G	$d_B = (\bar{X} - M_G)$	$(d_B)^2$
Group 1	5.29	5.77	−0.48	.23
Group 2	7.86	5.77	2.09	4.37
Group 3	5.00	5.77	−.77	.59
Group 4	6.29	5.77	.52	.27
Group 5	4.43	5.77	−1.34	1.80
				$\Sigma(d_B)^2 = 7.26$

The Raw Score Method

The definition method for calculating F is quite tedious to perform by hand on data containing decimal values and large N. Alternative calculation formulas for SS_B and SS_W that use raw scores rather than deviation scores make the computation much easier. Recall from equation 9.02 that $SS_T = SS_B + SS_W$. Because SS_T and SS_B are easier to calculate than SS_W, SS_W may be determined by the equation $SS_W = SS_T − SS_B$. Once SS_B and SS_W are found, the remainder of the calculation of F is the same as described in the definition method.

The computational formulas to calculate SS_B, SS_T, and SS_W are as follows:

$$SS_B = \frac{(\Sigma X_1)^2}{n_1} + \frac{(\Sigma X_2)^2}{n_2} + \frac{(\Sigma X_3)^2}{n_3} \ldots + \frac{(\Sigma X_k)^2}{n_k} - \frac{(\Sigma X_T)^2}{N} \quad (9.08)$$

$$SS_T = \Sigma X^2 - \frac{(\Sigma X_T)^2}{N} \qquad (9.09)$$

$$SS_W = SS_T - SS_B \qquad (9.10)$$

The raw score method requires that each individual raw data value (X) be squared and the sums of squares for each group be calculated. The raw data from table 9.1 are squared and summed in table 9.5.

Applying the data from tables 9.1 and 9.5 to equations 9.08, 9.09, and 9.10 yields

$$SS_B = \frac{37^2}{7} + \frac{55^2}{7} + \frac{35^2}{7} + \frac{44^2}{7} + \frac{31^2}{7} - \frac{202^2}{35}$$

$$SS_B = 1216.57 - 1,165.83 = 50.74$$

and

$$SS_T = 1,264 - \frac{202^2}{35} = 98.17$$

and

$$SS_W = 98.17 - 50.74 = 47.43.$$

These values differ slightly from the values calculated by the definition method, because the definition method requires rounding and squaring of rounded values. The values calculated by the raw score method are more accurate.

Table 9.5   Raw Data (X) Squared and Summed

$(X_1)^2$	$(X_2)^2$	$(X_3)^2$	$(X_4)^2$	$(X_5)^2$
16	25	25	64	25
25	49	16	16	16
36	81	36	36	9
49	64	25	64	16
16	81	25	25	36
36	49	36	36	16
25	100	16	49	25
$\Sigma(X_1)^2 = 203$	$\Sigma(X_2)^2 = 449$	$\Sigma(X_3)^2 = 179$	$\Sigma(X_4)^2 = 290$	$\Sigma(X_5)^2 = 143$

$\Sigma X^2 = 203 + 449 + 179 + 290 + 143 = 1,264$

The computations of degrees of freedom, mean square values, and F are the same in both methods:

$$df_B = 5 - 1 = 4$$

$$df_E = 35 - 5 = 30$$

$$MS_B = \frac{50.74}{4} = 12.69$$

$$MS_E = \frac{47.43}{30} = 1.58$$

$$F = \frac{12.69}{1.58} = 8.03$$

Determining the Significance of F

The significance of F is determined by referring to tables A.4, A.5, and A.6 in appendix A, which show the values of F for appropriate degrees of freedom between and within groups. In these tables df_B are read across the top of the table, and df_E are read down the left-hand side.

Table A.4 shows the values of F for $p = .10$. When $df_B = 4$ and $df_E = 30$, the critical F value is 2.14. Because the obtained F (8.03) exceeds this value, we look in table A.5, which lists the F values for $p = .05$. For $p = .05$, critical $F = 2.69$. The obtained F is still larger than critical F, so we look in table A.6, which has F values for $p = .01$. For $p = .01$, critical $F = 4.02$. The obtained F (8.03) also exceeds this critical F value, so the obtained F is declared significant at $p < .01$. This means that the odds are less than 1 in 100 that an F larger than 4.02 would be obtained by chance alone. Therefore, the level of confidence reached by the obtained F value (8.03) is greater than 99%. We conclude that the treatment had an effect on at least one of the groups and reject H_0.

The results of analysis of variance are usually reported in table form. Table 9.6 is an example of a typical ANOVA table.

Table 9.6   Tabular Report of Analysis of Variance

Source of variance	Sum of squares	df	Mean square	F	p
Between groups	50.74	4	12.69	8.03	<.01*
Within groups (error)	47.43	30	1.58		
Totals	98.17	34			

* This value is only approximate from table A.6. Computer programs can calculate the exact p value for a given F.

Post Hoc Tests

As we discussed earlier, a significant F alone does not specify which groups differ from one another. It only indicates that there are differences somewhere among the groups. To identify the groups that differ significantly from one another, a post hoc test must be performed.

A post hoc test is similar to a t test, except post hoc tests have a correction for familywise alpha errors built into them. Some are more conservative than others. Conservative means that the tests are less powerful because they require larger mean differences before significance can be declared. Conservative tests offer greater protection against type I errors, but they are more susceptible to type II errors.

Several post hoc tests may be applied to determine the location of group differences after a significant F has been found. Two of the most commonly used tests, **Scheffé's confidence interval (I)** and **Tukey's honestly significant difference (HSD)**, are described here. Scheffé permits all possible comparisons, while Tukey permits only pairwise comparisons. Tukey is easier to apply, and is appropriate in most research designs. For information on other post hoc tests, see Keppel (1991, p. 170–177) or other advanced statistical texts.

Scheffé's Confidence Interval (I)

The Scheffé test is the most conservative post hoc test. It should be used if all possible comparisons—that is, more than just pairwise comparisons—are to be made. Pairwise comparisons contrast one group mean against another. All possible comparisons include pairwise comparisons plus comparisons of combinations of groups to single groups or other combinations. For example, the average of two treatment effects may be compared with the average of two other treatment effects, or the average of several treatments may be compared to the control group. Scheffé places no restrictions on the number of comparisons that can be made.

Scheffé's confidence interval (I) permits us to compare the raw score means for any two groups or combinations of groups. The formula is

$$I = \sqrt{(k-1)(F_\alpha)\left(\frac{2\,MS_E}{n}\right)},\qquad(9.11)$$

where k = number of groups, F_α = the value of F from tables A.4 to A.6 for a given p value and the df_B and df_E values used in ANOVA, MS_E = mean square error from ANOVA, and n = size of the groups.

If groups are not equal in size, equation 9.11 may be modified as follows to accommodate any two groups with unequal values of n:

$$I = \sqrt{(k-1)(F_\alpha)(MS_E)\left(\frac{1}{n_1}+\frac{1}{n_2}\right)}.\qquad(9.12)$$

Because the ns in our example are equal, we apply equation 9.11 to the data from table 9.6. We compute I at $p = .01$ and $p = .05$ as follows:

$$I = \sqrt{(5-1)(4.02)\left(\frac{(2)(1.58)}{7}\right)} = 2.69, \quad p = .01$$

$$I = \sqrt{(5-1)(2.69)\left(\frac{(2)(1.58)}{7}\right)} = 2.20, \quad p = .05.$$

This interval size (I) is the raw score value by which any two means must differ to be considered significant. Constructing a mean difference table makes it easy to identify the groups that differ from one another.

Table 9.7 shows the differences between all pairwise combinations of means. Scheffé's I requires a mean difference of 2.20 for $p = .05$ and 2.69 for $p = .01$. The table identifies groups 2 and 1 as significantly different at $p < .05$, and groups 2 and 3, and 2 and 5 significantly different at $p < .01$.

The ordered values of the means provide additional insight.

Group	Mean
2	7.86
4	6.29
1	5.29
3	5.00
5 (control)	4.43

Table 9.7   Mean Difference Analysis

	Group 1	Group 2	Group 3	Group 4	Group 5
Group 1	0.00	2.57*	.29	1.00	0.86
Group 2		0.00	2.86**	1.57	3.43**
Group 3			0.00	1.29	0.57
Group 4				0.00	1.86
Group 5					0.00

Note: Values above are calculated by taking the absolute value of the difference between two means in table 9.1 (i.e., $|\bar{X}_1 - \bar{X}_2| = |5.29 - 7.86| = 2.57$).
* $p < .05$.
** $p < .01$.

It is clear that the treatment given to group 2 had a significant effect. Group 2 differs significantly from all other groups except group 4. Groups 1, 3, and 4 do not differ significantly from control.

An alternate method of identifying specific group mean differences using Scheffé's method may be developed by solving equation 9.12 for F. This results in the following:

$$F_{\text{Scheffé}} = \frac{(\bar{X}_1 - \bar{X}_2)^2}{(k-1)(MS_E)\left(\dfrac{1}{n_1} + \dfrac{1}{n_2}\right)}. \qquad (9.13)$$

$F_{\text{Scheffé}}$ is an F value calculated for any two specified groups (in the equation, $\bar{X}_1$ and $\bar{X}_2$ are specified, but any two means could be used). It is interpreted differently than I. I represents the raw score mean difference between any two groups that must be attained for declaration of significance. But $F_{\text{Scheffé}}$ is an actual F value that must be compared to tables A.4 to A.6 for df_B and df_E in the ANOVA analysis to determine if the two compared means differ significantly.

Equation 9.13 is sometimes used by computer programs to calculate the F values for all possible pairwise comparisons. If the computer does not also print the p values, the $F_{\text{Scheffé}}$ values produced by the computer must be compared with critical values in tables A.4 to A.6 to determine the significance of the difference between any two groups.

Tukey's Honestly Significant Difference (HSD)

Tukey's honestly significant difference (HSD) test, like I, calculates the minimum raw score mean difference that must be attained to declare significance between any two groups. But Tukey's test does not permit all possible comparisons; it only permits pairwise comparisons: Any single group mean may be compared to any other group mean. The formula for HSD is

$$HSD = q_{(k,df_E)} \sqrt{\frac{MS_E}{n}}, \qquad (9.14)$$

where q = a value from the Studentized range distribution (see tables A.7, A.8, and A.9 in appendix A) for k and df_E at a given level of confidence (note that k is used, not df_B), MS_E is the mean square error value from the ANOVA analysis, and n is the size of the groups.

Equation 9.14 assumes the ns in each group are equal. It may be modified to compare any two groups with unequal values of n as follows:

$$HSD = q_{(k,df_E)} \sqrt{\frac{MS_E}{2}\left(\frac{1}{n_1} + \frac{1}{n_2}\right)}. \qquad (9.15)$$

Because *n*s are equal in our example, equation 9.14 is applied for both $p = .01$ ($q = 5.05$) and $p = .05$ ($q = 4.10$) as follows:

$$HSD = 5.05 \sqrt{\frac{1.58}{7}} = 2.40, \ p = .01$$

$$HSD = 4.10 \sqrt{\frac{1.58}{7}} = 1.95, \ p = .05.$$

These values (2.40 and 1.95) represent the minimum raw score differences between any two means that may be declared significant.

Tukey, a more liberal test, confirms Scheffé but also finds that groups 2 and 1 differ at $p < .01$ (see table 9.7). The values at $p = .01$ and $p = .05$ are both lower in Tukey's *HSD* test than for Scheffé's *I*. This makes *HSD* more powerful (i.e., more likely to reject H_0) than *I*.

Because we started with the null hypothesis and are not making any comparisons other than pairwise (i.e., we are not interested in the combined mean of two or more groups compared to other combined means), Tukey's test is appropriate. Scheffé may be too conservative for this research design. Based on the analysis by Tukey, group 2 is significantly different from groups 1, 3, and 5 at $p < .01$. Figure 9.1 presents the results in bar graph form. The T symbol above each bar represents standard deviation.

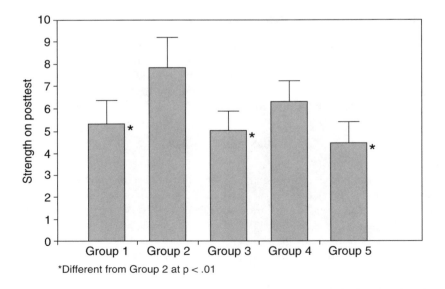

Figure 9.1   ANOVA on strength scores. Mean plus or minus standard deviation.

Concluding Statement Regarding Post Hoc Tests

When testing the null hypothesis with simple ANOVA, a researcher should first conduct an F test to determine if there are any differences among any of the groups and to determine the value of MS_E. If F is not significant, no further calculations are needed. If F is significant and comparisons other than pairwise comparisons are needed, Scheffé should be used to investigate differences among various combinations of groups. If F is significant, and only pairwise comparisons are to be made, Tukey should be used to contrast mean differences.

The Magnitude of the Treatment (Size of Effect)

A significant F is not the only factor considered when evaluating the importance of a finding. The researcher should also consider the size of the effect of the treatment. A significant F indicates the probability that the differences occurred by chance. ANOVA is actually a test of the reliability of the differences. When F is significant, the odds are good that similar mean differences would be found if the experiment were repeated. However, when N is large and MS_E is small, ANOVA may produce a significant F when the mean differences are so small that they have little or no practical value. Determining the size of the effect is a method of comparing treatment effects, independent of sample size.

For example, two groups of 200 subjects each participate in aerobic exercise programs for 4 days a week and 5 days a week. We may find that VO_2max values in milliliters per kilogram per minute are 47.6 and 49.1, respectively, and that the difference is significant at $p < .05$. Although this difference may be real, or not due to chance, it is probably not large enough (1.5 milliliters per kilogram per minute) to be worth the effort of an additional workout per week.

R^2

The most simple measure of the effects of treatment in ANOVA is R^2:

$$R^2 = \frac{SS_B}{SS_T}.$$

(9.16)

R^2 is the ratio of the variance due to treatment and the total variance. It produces a rough estimate of the size of the effect. For the data from the strength training experiment (table 9.6), $R^2 = 50.74 / 98.17 = .52$. This means that 52% of the total variance can be explained by the treatment effects. The remaining 48% is unexplained.

The value of R^2 will vary between 0.00 and 1.00, depending on the relative size of the treatment effects. It is a measure of the proportion of the total variance that is explained by the treatment effects. In this case, a little more than half of the total variance is explained. This is a fairly large proportion and confirms the conclusion that at least one treatment was effective in improving strength. In some statistics books (see Tabachnick & Fidell, 1989, p. 93), R^2 is called **eta squared.**

Omega Squared (ω^2)

Another method of determining the size of the effect is **omega squared** (ω^2):

$$\omega^2 = \frac{SS_B - (k-1)(MS_E)}{SS_T + MS_E}. \tag{9.17}$$

For the data in table 9.6, ω^2 is calculated as follows:

$$\omega^2 = \frac{50.74 - (5-1)(1.58)}{98.17 + 1.58} = .45.$$

Omega squared is a more accurate measure of effect size because it attempts to account for the unexplained variance (MS_E) and usually produces a smaller value than R^2. The value of $\omega^2 = .45$ is smaller than $R^2 = .52$, but it still indicates that a large proportion of the total variance may be attributed to treatment effects. This would still be considered a large effect size. Keppel (1991, p. 66) suggests that for ω^2 a value of .15 is large, .06 is medium, and .01 is small in the behavioral sciences.

The size of the effect is a relatively new concept in experimental design, but it is an important one to calculate and report. Thomas, Salazar, and Landers (1991, p. 344) state:

> Authors need to be convinced that they should report the magnitude of effects, as small differences can be easily declared significant based on some combination of small variances and large Ns (Thomas and Nelson, 1996), or the reverse can occur—large differences can be declared nonsignificant due to large variances and small Ns.

An Example From Leisure Studies/Recreation

Researchers in the effective use of leisure time have postulated that play activity may be used to reduce stress. It is proposed that play activity is most effective when it is perceived by the subject to be free play, or not directed by others. To test this hypothesis, Finney (1985) randomly divided male and female college-age subjects into four groups: high perceived control of the play experience (i.e., low structure), moderate perceived control, low perceived control, and a control group, who performed what they considered to be work, not play. The groups had 19, 20, 20, and 20 subjects, respectively.

All subjects performed a 30-minute stress-producing task—they worked 12 pages of math problems while listening to periodic bursts of 95 decibels of noise delivered through headphones. Upon completing this 30-minute stress period, the subjects engaged for 10 minutes in one of four play activities that varied in the amount of perceived control that the subjects had over their own play behavior. Following this play period, the subjects attempted to solve four geometric puzzles. The subjects were unaware that two of the puzzles were not solvable. Persistence on the two unsolvable puzzles (measured by time in total seconds spent on the two puzzles before giving up) was the dependent variable used to assess the effectiveness of the play period in reducing the stress created by the work task. Whereas H_1 was used to justify the study, the null hypothesis H_0 was tested statistically.

A simple ANOVA applied to the data yielded the information in tables 9.8 and 9.9. To identify specific mean differences, Finney applied $F_{Scheffé}$ (equation 9.13) to the comparisons between each group. This produced a matrix of F values contrasting each group with every other one. Table 9.10 demonstrates this analysis.

Table 9.8   Effects of Play on Reduction of Stress

Source of variance	Sum of squares	df	Mean square	F	p
Between groups	1,193,210.000	3	397,736.667	4.189	<0.01
Within groups (error)	7,121,448.000	75	94,952.640		
Totals	8,314,658.000	78			

Table 9.9   Group Means (Time in Seconds) and SD by Level of Structured Play

	High	Medium	Low	Control
Mean	705.90	466.30	427.90	385.05
Standard deviation	507.64	251.93	235.31	108.80

Note: Data are in total seconds.

Table 9.10 $F_{Scheffé}$ Values for Pairwise Comparisons

	High	Medium	Low	Control
High	0.000	1.964	2.643	3.521*
Medium		0.000	0.052	0.232
Low			0.000	0.064
Control				0.000

* $p < .05$.

The findings of $F_{Scheffé}$ declare that only the group with perceived high control of the play period differs from the control group at $p < .05$. (Table A.6 indicates that critical F for $df_B = 3$ and $df_E = 75$ is 4.13 at $p = .01$, and table A.5 indicates critical F is 2.76 at $p = .05$.)

Thinking that perhaps $F_{Scheffé}$ may be too conservative for this type of data, Finney then applied Tukey's post hoc test to analyze specific raw score mean differences (see table 9.11). Using equation 9.15, Finney found Tukey's HSD was 316.95 at $p = .01$ and 257.79 at $p = .05$. Tukey confirms Scheffé by indicating that the high group differs from the control group, but with Tukey, the difference is found to be significant at $p < .01$. The Tukey test also found that the high group differed from the low group at $p < .05$. But the low and medium groups still did not show significant differences between each other or from the control group.

The value of ω^2 was .1080, revealing a moderate effect size. About 11% of the differences between the groups can be attributed to the play treatment.

Table 9.11   Mean Difference Values for Pairwise Comparisons

	High	Medium	Low	Control
High	0.00	239.60	276.00*	320.85**
Medium		0.00	38.40	81.25
Low			0.00	42.85
Control				0.00

** $p < .01$; * $p < .05$.

Summary

ANOVA compares the means of three or more groups. When F is significant, it indicates that somewhere among the several means there is a significant difference. However, the specific mean differences are not identified. A post hoc test must be used to identify which means differ. After a significant F is found, some measure of the size of the effect of the treatment should be computed to ascertain the relative proportion of the differences that can be attributed to the treatment effects.

Remember that when two means differ significantly at some p value (for example, $p = .05$), the p value indicates the odds (5 in 100) that the means really are not different—that is, the odds that the null hypothesis is true. This is the same as stating that we are 95% sure that the null is false (95% level of confidence). We typically start with the assumption that there is no difference (H_0), and then we test that assumption. If we find significant F values, the null is rejected at the given level of confidence. We never really know if we are correct in accepting or rejecting the null hypothesis, we only know the odds, or probability, that we are correct in our conclusion.

Problems to Solve

Check your hand-calculated answers against computer-generated results.

1. A biomechanics researcher wanted to test whether good, average, and poor sprinters differed in their horizontal foot speed. She classified the sprinters into three groups based on their sprint times. The horizontal foot speed at touchdown in feet per second was then analyzed with the following results:

   Good 4, 5, 8, 6, 7, 6
   Average 7, 8, 9, 6, 7, 10
   Poor 10, 13, 12, 8, 11, 12

   A. What are the mean values for each group?
   B. What are the sum of squares values?
   C. What are the degrees of freedom and mean square values?
   D. What is the F value? Is it significant? If so, what is the p value?
   E. Set up a mean difference table.
   F. Apply Scheffé's I to the data. Do any of the means differ?
   G. Apply Tukey's HSD test. Do any of the means differ?
   H. Does Tukey differ from Scheffé in the interpretation of the mean difference table?
   I. What is the size of the effect? Is it small, moderate, or large?

2. A physical education teacher wanted to know the effects of various activity levels on body composition. She surveyed her physical education classes and categorized 30 students into five activity levels: inactive, semiactive, normal, active, and very active, based on the amount of daily exercise in which the students participated. She measured percent body fat using skinfold calipers on all subjects with the following results.

Inactive	Semiactive	Normal	Active	Very Active
30.2	29.4	22.9	17.6	10.9
29.6	17.6	25.4	13.4	13.7
35.2	26.4	19.6	20.3	12.8
19.1	25.3	18.7	19.6	14.7
26.3	22.5	21.8	15.1	9.3
22.4	28.6	24.9	10.7	12.7

Use a computer to solve this problem.

     A. What is the independent variable in this study? What is the dependent variable?

     B. What are the means and standard deviations for each group?

     C. Is there a significant difference among any of the groups? What is the probability that the null hypothesis is true?

     D. Create a mean difference table and indicate the significance of the differences among the groups.

See appendix C for answers to problems.

Key Words in This Chapter

Analysis of variance	Mean square
Familywise error rates	Scheffé's confidence interval *(I)*
Bonferroni adjustment	Tukey's honestly significant difference
Between-group variance	*(HSD)*
Within-group variance	Eta squared (R^2)
Total variance	Omega squared
Sum of squares	

Chapter 10

Analysis of Variance With Repeated Measures

An exercise physiologist wanted to know if there are significant differences among the heart rates of subjects who walk at 2, 4, and 6 miles an hour on a treadmill. All subjects walk on a treadmill at increasing speeds, and heart rate is measured for each subject at 2, 4, and 6 mph. The mean values were determined to be 85 beats per minute (bpm) at 2 mph, 120 bpm at 4 mph, and 150 bpm at 6 mph. Are these differences significant? To answer this question, the researcher must use a test that accounts for the fact that the mean values are not independent—there is a relationship among the scores because each set of scores was taken from the same person. What statistical test should be used?

One of the most common research designs in kinesiology involves measuring subjects before treatment (pretest) and then after treatment (posttest). A dependent t test with matched or correlated samples is used to analyze such data, because the same subjects are measured twice **(repeated measures).** In ANOVA, this type of design is referred to as a **within-subjects design.** When three or more tests are given—for example, if a pre-, mid-, and posttest are given with treatment before and after the midtest—ANOVA with repeated measures is needed to properly analyze the differences among the three tests.

The simple ANOVA described in chapter 9 assumes that the mean values are taken from independent groups that have no relationship to each other. In this independent group design, total variability is the sum of

- variability between people in the different groups **(interindividual variability),**
- variability within a person's scores **(intraindividual variability),**
- variability between groups due to treatment effects, and
- variability due to error (variability that is unexplained).

When there is only one group of subjects, but they are measured more than once, the data sets are dependent. The total variability for repeated measures on one group is expected to be less than if the scores came from different groups of people (that is, were independent), because interindividual variability has been eliminated by using a single group. This tends to reduce the MS_E term in the denominator of F in a manner similar to the correction made to the standard error of the difference in the t test (equation 8.09).

The advantage of the repeated measures design is that the same subjects are measured repeatedly. In this way, the subjects serve as their own control. If all other relevant factors have been controlled, any differences observed between the means must be due to (a) the treatment, (b) variations within the subjects (intraindividual variability), or (c) error (unexplained variability). Variability between subjects (interindividual variability) is no longer a factor. The formulas previously presented for simple ANOVA are modified to account for repeated measures.

Assumptions in Repeated Measures ANOVA

Except for independence of samples, the assumptions for simple ANOVA, or between-subjects designs, discussed in chapter 9 also hold true for repeated measures ANOVA, or within-subjects designs. But with repeated measures designs, we must also consider the relationships among the repeated measures. Repeated measures ANOVA must meet the assumption of **sphericity,** sometimes referred to as compound symmetry. Sphericity requires that the repeated measures demonstrate homogeneity of variance and homogeneity of covariance. **Homogeneity of covariance** means that the relationships, or correlations, on the dependent variable among all of the three or more repeated measures are equal. When only two repeated measures are employed (such as pre-post measures for a *t* test), this assumption is not applicable, because there is only one correlation coefficient that can be calculated, the correlation between pre- and posttest scores. Methods of dealing with violations of sphericity will be presented later in this chapter.

[handwritten margin note: Key point = sphericity. assumption not applicable an pre-post designs b/c only one btw pre and post.]

Calculating Repeated Measures ANOVA

To demonstrate how to calculate ANOVA with repeated measures, we will analyze a hypothetical study. A graduate student in biomechanics was interested in the decrease in balance ability that bicycle racers experience as their fatigue from the race increases. To measure this, the researcher placed a racing bicycle on a roller ergometer. The bicycle's front and rear wheels were placed on the tops of cylindrical rollers so that as the wheels turned, the rollers turned and the rider was able to ride in place (figure 10.1).

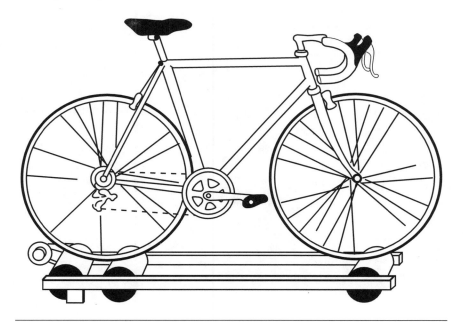

Figure 10.1   Roller ergometer.

A 4-inch-wide stripe was painted in the middle of the front roller, and the rider was required to place the front wheel on the stripe. The rear rollers were connected to a braking system to provide resistance to the rear wheel of the bike. Balance was indicated by wobble in the front wheel and was measured by counting the number of times per minute that the front wheel of the bike strayed off the stripe.

As resistance on the rear wheels increased, physiological fatigue increased, and it became more and more difficult to maintain the front wheel on the stripe. Subjects rode the bicycle for 15 minutes, divided into five 3-minute periods for the purpose of collecting data. Data were collected on the number of balance errors during the last minute of each 3-minute period, and resistance was increased at the end of each 3-minute period. In this design, the dependent variable is balance errors and the independent variable is increase in resistance.

Table 10.1 presents the raw data in columns and rows. The data (in errors per minute) for the subjects ($N = 10$) are in the rows, and the data for the trials ($k = 5$ repeated measures) are in the columns. The sum of rows (ΣR) is the total score for each subject over all five trials, and the sum of rows squared ($\Sigma R)^2$ is presented at the right of the table. The sum of each column (ΣC) is the total for all 10 subjects on a given trial; ΣX_T is presented at the bottom of the table.

Table 10.1   Raw Data: Balance Errors per Minute

Subject	Minute 3	Minute 6	Minute 9	Minute 12	Minute 15	ΣR	$(\Sigma R)^2$
1	7	7	23	36	70	143	20,449
2	12	22	26	26	20	106	11,236
3	11	6	9	31	30	87	7,569
4	10	18	16	40	25	109	11,881
5	6	12	9	28	37	92	8,464
6	13	21	30	55	65	184	33,856
7	5	0	2	10	11	28	784
8	15	18	22	37	42	134	17,956
9	0	2	0	16	11	29	841
10	6	8	27	32	54	127	16,129
ΣC	85	114	164	311	365	1,039	129,165
Mean =	8.5	11.4	16.4	31.1	36.5		

$\Sigma X_T = 85 + 114 + 164 + 311 + 365 = 1,039$

Steps in Calculation

1. Arrange the raw data (X) in tabular form, placing the data for subjects in rows (R), and repeated measures in columns (C) as shown in table 10.1.

2. Calculate the row totals ($\Sigma R_1, \Sigma R_2, \Sigma R_3, \ldots$), the column totals ($\Sigma C_1, \Sigma C_2, \Sigma C_3, \ldots$), and the grand total: $\Sigma X_T = 1,039$.

3. Square each row total, $(\Sigma R)^2$, and sum these values: $\Sigma (\Sigma R)^2 = 129,165$.

4. Compute the mean values for columns.

5. Square each raw score (X^2) and calculate the squared column totals and grand total: $\Sigma X^2 = 34,947$, as shown in table 10.2.

6. Compute the sum of squares between columns (SS_c). This is the variability due to treatment effects.

$$SS_c = \frac{(\Sigma C_1)^2 + (\Sigma C_2)^2 + (\Sigma C_3)^2 + \ldots + (\Sigma C_k)^2}{N} - \frac{(\Sigma X_T)^2}{(N)(k)}. \quad (10.01)$$

Therefore, in the bicycle problem:

$$SS_c = \frac{85^2 + 114^2 + 164^2 + 311^2 + 365^2}{10} - \frac{1,039^2}{(10)(5)},$$

$$SS_c = 27,706.3 - 21,590.42 = 6,115.88.$$

7. Calculate the sum of squares between rows (SS_R). This is the variability due to differences among the subjects.

$$SS_R = \frac{\Sigma(\Sigma R)^2}{k} - \frac{(\Sigma X_T)^2}{(N)(k)} \quad (10.02)$$

Table 10.2 Raw Data Squared

Subject	Minute 3	Minute 6	Minute 9	Minute 12	Minute 15
1	49	49	529	1,296	4,900
2	144	484	676	676	400
3	121	36	81	961	900
4	100	324	256	1,600	625
5	36	144	81	784	1,369
6	169	441	900	3,025	4,225
7	25	0	4	100	121
8	225	324	484	1,369	1,764
9	0	4	0	256	121
10	36	64	729	1,024	2,916
Total	905	1,870	3,740	11,091	17,341

$\Sigma X^2 = 905 + 1,870 + 3,740 + 11,091 + 17,341 = 34,947$

Therefore, in the bicycle problem:

$$SS_R = \frac{129,165}{5} - \frac{1,039^2}{(10)(5)},$$

$$SS_R = 25,833.0 - 21,590.42 = 4,242.58.$$

8. Calculate the total sum of squares (SS_T). This is the variability due to subjects (rows), treatment (columns), and unexplained residual variability (error).

$$SS_T = \Sigma X^2 - \frac{(\Sigma X_T)^2}{(N)(k)} \qquad (10.03)$$

In the bicycle problem:

$$SS_T = 34,947 - \frac{1,039^2}{(10)(5)},$$

$$SS_T = 34,947 - 21,590.42 = 13,356.58.$$

9. Calculate sum of squares due to error (SS_E). This is the unexplained variance that will be used in the denominator of F. It is the amount of variance that can be attributed to chance. Because variability from columns (treatments), rows (subjects), and error must equal total variability, we may find SS_E by subtraction:

$$SS_E = SS_T - SS_C - SS_R. \qquad (10.04)$$

Therefore:

$$SS_E = 13,356.58 - 6,115.88 - 4,242.58 = 2,998.12.$$

10. Calculate the degrees of freedom for each source of variance.
Columns (treatment)

$$df_C = k - 1, \qquad (10.05)$$
$$df_C = 5 - 1 = 4.$$

Rows (subjects)

$$df_R = N - 1, \qquad (10.06)$$
$$df_R = 10 - 1 = 9.$$

Error

$$df_E = (k - 1)(N - 1), \qquad (10.07)$$
$$df_E = (5 - 1)(10 - 1) = 36.$$

Total

$$df_T = (N)(k) - 1,$$ (10.08)

$$df_T = (10)(5) - 1 = 49.$$

11. Calculate the mean square for each source of variance.
Mean square columns (trials)

$$MS_C = \frac{SS_C}{df_C},$$ (10.09)

$$MS_C = \frac{6,115.88}{4} = 1,528.97.$$

Mean square rows (subjects)

$$MS_R = \frac{SS_R}{df_R},$$ (10.10)

$$MS_R = \frac{4,242.58}{9} = 471.40.$$

Mean square error

$$MS_E = \frac{SS_E}{df_E},$$ (10.11)

$$MS_E = \frac{2,998.12}{36} = 83.28.$$

12. Calculate F for columns (treatment)

$$F_C = \frac{MS_C}{MS_E},$$ (10.12)

$$F_C = \frac{1,528.97}{83.28} = 18.36.$$

It is also possible to calculate an F value for subjects, or rows ($F_R = MS_R / MS_E$). But this value is not of interest at this time because it simply represents a test of the variability among the subjects. This F_R value is not important to the research question being considered: Does an increase in fatigue result in an increase in mean balance errors across trials (columns)? We will use F_R later in this chapter when intraclass reliability is discussed.

Determining the Significance of F

To determine the significance of F for columns, we look in tables A.4, A.5, and A.6 of appendix A. The degrees of freedom for columns df_C is the same as df_B from chapter 9 because the columns represent variability between trials. The degrees of freedom for error df_E is used in the same way as df_E in chapter 9, to measure within-group variability. Table A.6 shows that for df (4,36) an F of 4.02 is needed to reach $p = .01$. Because our obtained F (18.36) easily exceeds 4.02, we conclude that differences do exist somewhere among the mean values for the five trials at $p < .01$.

Correcting for Violations of the Assumption of Sphericity

One assumption of repeated measures ANOVA is that the variance of the several repeated measures, or trials, is equal (homogeneity of variance), and the correlations among all combinations of trials (homogeneity of covariance) are equal. This is called the assumption of sphericity.

If the assumption of sphericity is not true, the probability of making a type I error increases. Methods of correcting for a violation of the assumption have been suggested by Greenhouse-Geisser and Huynh-Feldt. Both correction procedures modify the df values for columns (treatment) and error.

The Greenhouse-Geisser Adjustment

The **Greenhouse-Geisser (GG) adjustment** consists of dividing the df_C and df_E values by $k - 1$, the number of repeated measures minus 1. Note that $k - 1$ is the same as the value for df_C (equation 10.05). Hence, with the GG adjustment, the value df_C becomes 1: $df_C / (k - 1) = 1$. Degrees of freedom for error are adjusted by the same method: $df_E = df_E / (k - 1)$. In the bicycle research example, the adjusted df_E is 9: $36 / (5 - 1) = 9$. We then reevaluate F using tables A.4, A.5, and A.6 and the adjusted df (1, 9). For df (1, 9), table A.6 indicates that F must equal 10.56 at $p = .01$. Because our obtained value (18.36) exceeds 10.56, we may still conclude that the means of two or more trials are significantly different at $p < .01$.

This application of the GG adjustment assumes maximum violation of the assumption of sphericity. Consequently, when the violation is minimal, this adjustment of df may be too severe, possibly resulting in a type II error; accepting the null hypothesis when it is false.

The Huynh-Feldt Adjustment

An alternate method, the **Huynh-Feldt (HF) adjustment,** attempts to correct for only the amount of violation that has occurred. In the HF adjustment, df_C and df_E are multiplied by a value (epsilon, ϵ) that ranges from 0.00 for maximum violation to 1.00 for no violation. As a general rule, if $\epsilon \geq .75$, the violation is considered to be insignificant. For example, if $\epsilon = .43$, $df_C = 4 \times .43 = 1.72$ and $df_E = 36 \times .43 = 15.48$. For $df (1.72, 15.48)$, table A.6 indicates an F value of 6.36 is needed to reach $p = .01$. Because the obtained F (18.36) exceeds 6.36, we are better than 99% confident in declaring F to be significant at $p < .01$ even if sphericity is violated where $\epsilon = .43$.

But even though F is still significant, these adjustments reduce the confidence we can place in our conclusion that the differences among the trial means are significant. If the obtained p value is close to the rejection level of $p = .05$ (suppose we obtain $p = .04$), and the adjustment raises it to .06, H_0 must be accepted.

Calculation of epsilon is a complicated procedure and will not be discussed here. Winer, Brown, and Michels (1991, p. 251) present procedures for calculating epsilon. Most advanced computer programs (BMDP, SPSS, SAS, etc.) test for violations of the assumption of sphericity and provide both the Greenhouse-Geisser and the Huynh-Feldt adjustments, including the epsilon values.

A strategy for determining the significance of F (when sphericity must be assumed) has been suggested by Keppel (1991, p. 353). A modification of Keppel's strategy follows.

Evaluate F with the GG adjustment (the most conservative condition):

- If F with GG adjustment is significant, reject H_0.
- If F with the GG adjustment is not significant, evaluate F with no adjustment.
- If F with no adjustment is not significant (the most liberal condition), accept H_0.
- If F with GG adjustment is not significant, but F with no adjustment is significant, use HF adjustment (a moderate condition) to make the final determination.

An alternate solution if violations are severe is to use MANOVA (see chapter 12) with the repeated measures designated as multiple dependent variables. Under these conditions, the assumption of sphericity is not required. This approach is less powerful and thus provides better protection against type I errors. Further discussion of sphericity may be found in Dixon (1990, p. 504), Keppel (1991, p. 351), Schutz and Gessaroli (1987), Tabachnick and Fidell (1989, pp. 46 and 339), and Winer, Brown, and Michels (1991, p. 251). The ANOVA results for the bicycle test are summarized in table 10.3, as calculated by computer.

Table 10.3   ANOVA: Bicycle Errors

Source	Sum of squares	df	Mean square	F	p
Treatment (columns)	6,115.88	4	1,528.97	18.36	0.0000
Error	2,998.12	36	83.28		

Note. GG $\epsilon = .37$; $p < .0003$. HF $\epsilon = .43$; $p < .0001$

Post Hoc Tests

In this study, the researcher was interested in how much fatigue could be tolerated before a significant decrease in balance ability occurred. As the racer pumps up a hill, fatigue increases and balance decreases. It may be wise for the racer to rest at the top of a long hill, or at least proceed down the other side carefully, because balance may be impaired, and a serious fall could occur. But at what point does balance ability decrease significantly? Tukey's post hoc test of honestly significant differences is used to answer this question.

In chapter 9, Tukey's test (equation 9.14) was described as a method of making all pairwise mean comparisons. Tables A.8 and A.9 in appendix A list the values for q at k (5) and df_E (36) as 4.10 at $p = .05$ and 5.05 at $p = .01$. Note: In repeated measures designs, k is the number of repeated measures, and Ns are always equal. Therefore,

$$HSD_{.01} = 5.05 \sqrt{\frac{83.28}{10}} = 14.57.$$

Any two column means that differ by 14.57 balance errors per minute or more are determined to be significantly different at $p \leq .01$. A similar analysis could be made at $p \leq .05$ by using $q = 4.10$.

A graph of the column means indicating the position of the significant differences is presented in figure 10.2. Minutes 3, 6, and 9 do not significantly differ from one another, and minute 12 does not differ from minute 15 (i.e., they are less than

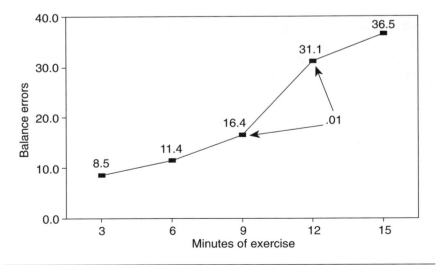

Figure 10.2   Mean balance errors over 15 minutes of exercise.

14.57 points apart). But minutes 9 and 12 differ by 14.7 (31.1 − 16.4 = 14.7) points. Therefore, between minutes 9 and 12, a significant increase ($p < .01$) in errors committed is observed. Because errors increase with each subsequent trial, any comparison of trials before minute 9 with minute 12 or 15 must also be significant.

The researcher concluded that the increase in work load between minutes 9 and 12 is sufficient to significantly increase balance errors. If this work load can be related to heart rate, racers may be counseled to measure their heart rate after pumping up a long hill. If heart rate at the top of the hill is equal to or higher than a level equivalent to the work load at minute 9, the rider would be wise to proceed down the other side carefully.

Interpreting the Results

The F for treatment (in the bicycle example) indicates that significant differences exist somewhere among the mean values of the trials (columns) at $p < .01$. The GG and HF adjustments do not reduce the p value below our rejection level. Therefore, we conclude that F is significant at better than $p = .01$ even after adjustment of df for possible violations of the assumption of sphericity. This finding rejects the null hypothesis and confirms the research hypothesis that balance errors increase as fatigue increases. Tukey's HSD post hoc analysis reveals that a significant difference in number of errors occurred between minutes 9 and 12. These hypothetical results suggest that bicycle riders are less able to maintain balance when they are fatigued.

An Example From Leisure Studies/Recreation

A graduate student in leisure studies postulated that students' conservation awareness would increase by their participating in a semester-long backpacking class at the university. A questionnaire to measure conservation awareness was developed and found to be valid and reliable. It was administered to 12 students at the beginning (pre), at midterm (mid), and after completion of the class (post). Means and standard deviations for the three measures are reported in table 10.4.

Table 10.4   Repeated Measures of Conservation Awareness

	Pre	Mid	Post
Mean	58.6	58.8	62.6
Standard deviation	9.3	8.0	6.4

Repeated measures ANOVA produced an F value of 1.94 among the three means. For df (2, 22), table A.4 in appendix A requires a value of 2.56 to reach $p = .10$. Because the obtained F value was less than the lowest critical value from table A.4, the graduate student accepted the null hypothesis and concluded that there was no difference among the three means. Conservation awareness, as measured by the survey, was not affected by participation in the class. Because F was not significant, sphericity was not an issue, and no further analysis with post hoc tests was needed.

Intraclass Reliability

In chapter 7, the concept of reliability was briefly discussed. Reliability is the repeatability, or consistency, in the repetition of a measurement process.

Pearson's correlation coefficient is sometimes incorrectly used to compute test-retest reliability. Correlation compares deviations (fluctuations in subjects' scores) from the mean on two measurements, but it is not sensitive to changes in the means of the scores. If the fluctuations in the scores from the first to the second test all occur in a systematic manner (i.e., they all go up or down by the same amount), the order and deviations of the scores on the two tests will be the same (i.e., the correlation coefficient will be high), but the two means may differ significantly. In such cases, Pearson's r may lead us to erroneously conclude that the scores are highly reliable, when in fact they are changing considerably. Pearson's r is unable to detect the change in the means.

Another limitation of correlation is that it may only be applied to two sets of data at one time, but we may be interested in the reliability over several measures. **Intraclass reliability** solves both of these objections: It may be used on two or more measures simultaneously, and it is sensitive to changes in both the order and the magnitude (mean differences) of the repeated values. It is called intraclass because it is designed to analyze repeated measures data on the same variable. Interclass (i.e., Pearson) is appropriate for analysis of two different variables. Pearson's correlation indicates the maintenance of relative order between two sets of data, which may be based on different measurement scales, and is therefore appropriate for interclass analyses. In fact, it is often used in this manner to assess validity. Under these conditions we are not concerned about differences in mean values because the units of measurement may differ (i.e., time vs. distance).

However, when determining reliability, researchers are usually interested in changes in both the order of the values and their magnitude. ANOVA with repeated measures may be used to assess reliability with sensitivity to order and magnitude combined.

To determine the reliability of the measurement of balance errors on the bicycle treadmill described in this chapter, the researcher collected balance error data on 5 pilot subjects (not the same ones as in the study) for 15 minutes (measuring every 3 minutes) with no resistance on the rear rollers. Data from the 5 pilot subjects are shown in table 10.5.

Repeated measures analysis of variance by computer produced the results in table 10.6. The variability among the subjects (rows) is included in this table. The high F value (43.63, $p < .0000$) for rows indicates that the subjects were significantly different in the number of errors they made across all trials. This is expected because not all subjects are of equal riding ability. The nonsignificant F for treatment indicates that there are no significant differences among the means of the five trials. This is not surprising because the treatment conditions for all of the trials remained the same (i.e., no increase in resistance).

To calculate intraclass reliability, we must calculate a mean square value that represents the sum of changes in the mean (column, or treatment effects) and error (MS_{C+E}):

$$MS_{C+E} = \frac{SS_C + SS_E}{df_C + df_E}.$$ (10.13)

Table 10.5   Reliability Evaluation on Pilot Subjects

Subject	Minute 3	Minute 6	Minute 9	Minute 12	Minute 15
1	10	12	11	9	10
2	14	15	13	15	16
3	6	5	6	7	8
4	12	10	11	13	12
5	8	9	7	8	9
Mean =	10.0	10.2	9.6	10.4	11.0

Table 10.6   ANOVA: Bicycle Errors on Pilot Data

Source	Sum of squares	df	Mean square	F	p
Treatment (columns)	5.36	4	1.34	1.18	N.S.
Subjects (rows)	198.96	4	49.74	43.63	.0000
$T \times S$ (error)	18.24	16	1.14		
Total	222.56	24			

Using the data from table 10.6, we calculate MS_{C+E} as follows:

$$MS_{C+E} = \frac{5.36 + 18.24}{4 + 16} = 1.18.$$

Intraclass reliability (R_1) is then determined:

$$R_1 = \frac{MS_R - MS_{C+E}}{MS_R}. \tag{10.14}$$

Using the data from table 10.6:

$$R_1 = \frac{49.74 - 1.18}{49.74} = .976.$$

The reliability, or consistency of subjects across trials (R_1), of the bicycle data, including the effects of changes in the mean values, is .976. This is a very high reliability value. As a general rule, R_1 values above .90 are considered high, from .80 to .89 moderate, and below .80 questionable for physiological data. In the behavioral sciences, values between .70 and .80 may be considered acceptable, depending on the type of measurement instrument involved.

Because the mean values do not change significantly (F for columns, or treatment, is not significant) and R_1 (.976) is high positive, we conclude that the subjects were able to ride with little variation in their performance minute by minute. This confirms that the bicycle treadmill instrument designed to measure balance errors is reliable.

Occasionally, we want to measure reliability without considering changes in the mean values over trials. If we ignore the variance due to changes in the mean values (columns), we can compute the variance due to order (rows) alone among the subjects over the several trials. This value is designated as R_2 and is sometimes referred to as coefficient alpha. To compute R_2, the value for MS_E is substituted for MS_{C+E} in equation 10.14:

$$R_2 = \frac{MS_R - MS_E}{MS_R}. \tag{10.15}$$

For the pilot data from table 10.6, the value of R_2 is computed as follows:

$$R_2 = \frac{49.74 - 1.14}{49.74} = .977.$$

When the mean values do not change, R_2 will not deviate much from R_1. The value of R_2 (.977) differs little from R_1 (.976) because the means are very similar over the five trials. It is appropriate to use R_1 for assessing intraclass reliability including order and magnitude, whereas R_2 is appropriate when changes in the means are to be ignored.

Interpreting Intraclass Reliability

If both the order of individual subject scores and the mean values on subsequent measures do not change much, both R_1 and R_2 will be high. If both order and mean values change, both R_1 and R_2 will decrease, but R_1 will be reduced the most.

To illustrate this concept, let us analyze the bicycle data when the resistance was increased over trials. If we apply intraclass reliability analysis to the bicycle error data from table 10.1, $R_1 = .517$ and $R_2 = .823$. This difference reflects the fact that R_1 is sensitive to mean value changes over trials but R_2 is not. Figure 10.2 confirms that the mean values do indeed change dramatically. When we factor out the mean differences over trials by using R_2, the reliability increases to .823.

The proper method of determining reliability is to calculate it on pilot data, then confirm it on the experimental data. If the pilot data does not demonstrate acceptable reliability, the techniques used to measure the variable of interest in the study must be revised until acceptable reliability is obtained. Without evidence that the measurements are reliable in a pilot study, we cannot be confident that the data obtained in the actual study will be correct.

Summary

Repeated measures ANOVA, one of the most common designs used in kinesiological research, permits the researcher to determine the significance of mean differences measured on the same subjects over repeated trials. It is often used in testing designs where subjects are measured before, during, and after treatment or where subjects are measured repeatedly over time, as when data is collected minute by minute on a treadmill or bicycle ergometer. It produces an F value that can be evaluated to determine if there are any significant differences among the mean values of the various trials. When F is significant, a post hoc test must be applied to determine the specific trial means that differ from one another.

Repeated measures ANOVA must meet the additional assumption of sphericity; that is, the variances among the trials must be approximately equal and the correlations among the trials must be approximately equal. When this assumption is not met, adjustments to the df and the p values obtained must be made with either the Greenhouse-Geisser or the Huynh-Feldt correction techniques.

Problems to Solve

Check your hand-calculated answers against a computer-generated result.

1. An exercise physiologist measured VO_2 (ml/kg/min) for 5 subjects on the treadmill in a graded exercise test (GXT). After 3 minutes of level walking for warm-up, the work load was increased every 2 minutes for 10 minutes. Five repeated measures of VO_2 were taken, one at the end of each 2-minute interval, and the following data were obtained:

Experimental Data

Subject	Minute 2	Minute 4	Minute 6	Minute 8	Minute 10
1	20	24	27	31	35
2	17	18	21	28	31
3	19	25	29	36	41
4	12	17	21	27	32
5	15	20	27	31	38

A. What are the mean values for each measured minute?
B. Calculate the sums of squares for columns (minutes), rows (subjects), and error.
C. Calculate df and mean square for columns (minutes), rows (subjects), and error.
D. Calculate F and determine the significance of F.
E. Set up an ANOVA table to present the results of this problem.
F. Is there a significant difference between any of the minutes?
G. What is Tukey's HSD for minutes? Which of the minutes differ?
H. Prior to the actual experiment, the researcher collected pilot data with no increase in work load over time. Are the pilot data reliable? What is R_1? What is R_2?

Pilot Data

Subject	Minute 2	Minute 4	Minute 6	Minute 8	Minute 10
1	20	21	19	21	24
2	17	18	19	20	19
3	19	18	20	21	20
4	12	13	13	14	12
5	15	16	15	16	17

2. A biomechanist compared four commercial abdominal exercise machines with abdominal crunches done without equipment (control). To make the comparison, he attached EMG electrodes to the upper rectus abdominus muscle in 10 male subjects and recorded the electrical activity (in millivolts) during exercise. Following are the results.

Subject	Machine 1	Machine 2	Machine 3	Machine 5	Control
1	920	913	1005	811	905
2	566	580	767	833	833
3	328	293	300	290	367
4	568	555	637	484	557
5	248	267	167	312	331
6	815	606	607	462	459
7	313	264	252	295	314
8	538	502	399	354	506
9	1475	1563	1174	1722	1581
10	465	434	375	300	352

Data courtesy of Dr. William Whiting, Director, Biomechanics Laboratory, Department of Kinesiology, California State University Northridge.

A. What are the means and standard deviations for each machine and control?
B. Did any of the machines elicit more EMG activity in the upper rectus abdominus muscle than control conditions?
C. What is the probability that the null hypothesis is true in this experiment?
D. Is it necessary to perform a post hoc test in this experiment?

See appendix C for answers to problems.

Key Words in This Chapter

Repeated measures	Homogeneity of covariance
Within-subjects design	Greenhouse-Geisser adjustment
Interindividual variability	Huynh-Feldt adjustment
Intraindividual variability	Intraclass reliability
Sphericity	

Chapter 11

Factorial Analysis of Variance

Imagine that a researcher wants to study the effect of practice time on learning a novel task of throwing accuracy. The researcher designs an experiment in which college students practice for 1, 3, or 5 days a week, 20 minutes a day, for 6 weeks. The researcher also wants to know if subjects with athletic experience (varsity athletes) benefit differently from the practice conditions than do students with no experience in athletic competition. Nine skilled athletes and nine students with no competitive experience are selected and assigned to the three practice conditions. After 6 weeks, all data have been collected. How can they be analyzed?

In chapters 9 and 10, analysis of variance was conducted on one factor. In chapter 9, the effects of differing exercise regimens (one factor with five different groups of subjects) on strength were considered. In chapter 10, the effect of physiological fatigue (one factor with five repeated measures on the same subjects) on bicycle balance skill was considered.

However, researchers often want to know the simultaneous effects of several **factors** on the dependent variable. For example, one could study the effects of five different exercise regimens (five levels of factor A) and gender (two levels of factor B) on strength. This design would answer the question: Do different exercise regimens have the same effect on men as they do on women for development of strength? This type of **factorial ANOVA** is often referred to as a between-between design because it analyzes differences on the dependent variable **between** different groups of subjects assigned to each of the five levels of factor A and between males and females (two levels of factor B) simultaneously.

Studies may also involve a **within** factor, where subjects are measured repeatedly. Later in this chapter we will consider a design where three groups of subjects learning under different conditions (factor A—between) are measured three times (factor B—within) on a novel motor skill. This design would be classified as between-within, between groups and within trials.

It is also possible to have a within-within design. For example, the same subjects are tested on different modes of exercise (e.g., treadmill vs. cycle ergometer—factor A—within) and measured on VO_2 in milliliters per kilogram body weight per minute (dependent variable) repeatedly minute by minute (factor B—within) until exhaustion. In this design, the same subjects are measured both within modes and within minutes.

In kinesiology, we are frequently interested in the effects of various treatments, training, teaching methods, or other factors as they impact different subgroups of the population such as males vs. females, athletes vs. nonathletes, young vs. old, fit vs. unfit, and so forth. In these cases, a factorial analysis of variance may be employed to determine the simultaneous effects of two or more factors on a dependent variable.

The designs of these studies are sometimes identified by the number of factors and the levels of each factor being studied. If two factors, such as levels of intensity of training regimens (low, medium, high) and levels of gender (male, female), were applied to strength development (the dependent variable), the design would be 3×2 (three levels of factor A combined with two levels of factor B). If levels of the factor of age (15–30 years, 31–50 years, and 51–70 years) were added to this study, the design would become $3 \times 2 \times 3$ (training $\times$ gender $\times$ age). Although it is theoretically possible to have any number of factors, usually no more than three are employed.

Factorial analysis of variance produces an F value for each factor and an F value for the interaction of the factors. **Interaction** is the combined effect of the factors on the dependent variable. Two factors interact when the differences between the mean values on one factor depend upon the level of the other factor. It may be thought of as a measure of whether the lines describing the effects of the two factors are parallel (see figures 11.1 to 11.6). If the slopes of the lines are not significantly different (i.e., the lines are parallel), then interaction is not significant. Interaction would not be significant if it were found that training

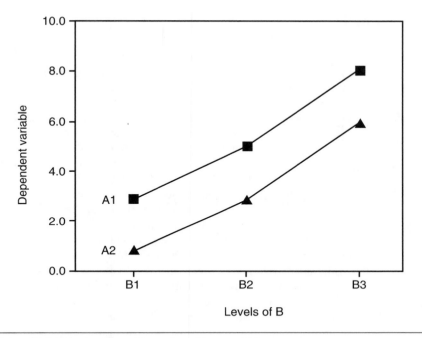

Figure 11.1   No significant interaction—lines parallel.

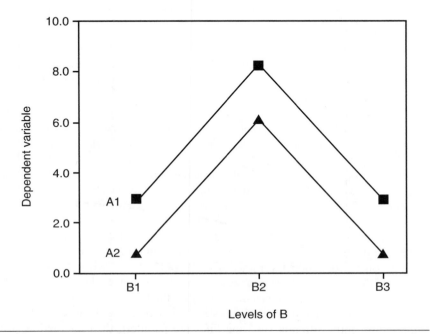

Figure 11.2   No significant interaction—lines parallel.

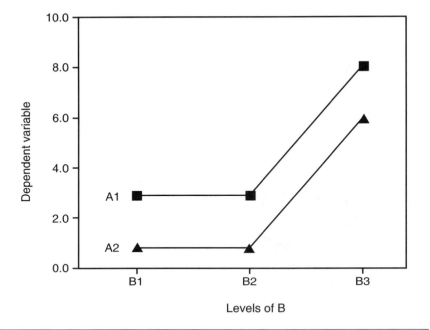

Figure 11.3   No significant interaction—lines parallel.

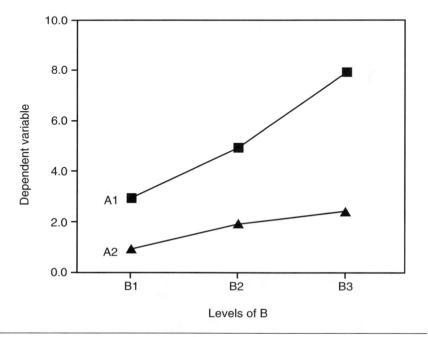

Figure 11.4   Significant interaction—lines not parallel.

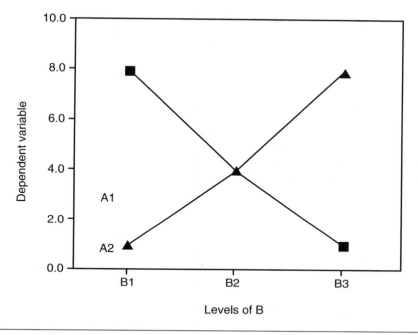

Figure 11.5   Significant interaction—lines not parallel.

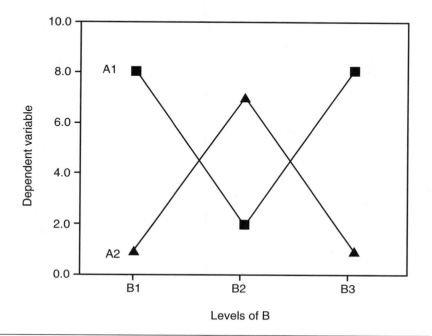

Figure 11.6   Significant interaction—lines not parallel.

regimens affect men and women in the same way, but it would be significant if it were found that the training regimens affect men differently than they do women. Further discussion of interaction will follow later in this chapter.

A Between-Between Example

To demonstrate between-between factorial analysis of variance, a hypothetical study of learning motor skills will be described. The data are fabricated and are intentionally simple so that the concepts may be easily understood and followed throughout the analysis without the confounding effect of complicated data. The assumptions for between-between designs are the same as for a simple ANOVA as described in chapter 9.

In this design, the researcher wanted to study the effect of 1 day, 3 days, or 5 days per week of practice (20 minutes per day) on the learning of a novel task of nondominant arm throwing accuracy (the dependent variable). In addition, the researcher wanted to know if subjects with athletic experience (college-level varsity athletes) benefit differently from the various practice conditions than do college students who have never had experience in athletic competition.

This between-between design (3 practice conditions $\times$ 2 groups of athletic experience) with equal numbers of subjects in each group is presented for demonstration purposes. However, more complicated designs with additional factors and unequal ns in each cell could be analyzed. Factorial ANOVA is typically performed by a computer that can deal with any number of factors, any number of levels, and with equal or unequal ns. When the concepts are understood for a simple design, more complicated designs will be easy to comprehend and the computer can do the work. The actual formulas for calculating the results will not be presented here. It is assumed that the student will have a computer that can perform this analysis. The emphasis in this text will be on understanding the concepts of the analysis and how to interpret the results.

Nine highly skilled varsity athletes from a basketball team were selected as the experienced group, and nine college students with no high school or college competitive sport experience were randomly selected from the student body and identified as the no experience group. Athletic experience vs. no experience was identified as factor A.

Three subjects in each group were then randomly assigned to one of the three practice conditions (1 day/week, 3 days/week, or 5 days/week of practice for 20 minutes per day). Number of days of practice was identified as factor B. Subject's scores on the throwing task at the end of 6 weeks were recorded as the dependent variable. Table 11.1 presents the raw data at the end of 6 weeks.

Table 11.1 Raw Data and Factor Sums

| | | Factor A (experience) | | |
		A₁ (athletes)	A₂ (nonathletes)	Sums of B
	B₁ (1/wk)	1	2	
		2	2	9
		1	1	
Factor B	B₂ (3/wk)	3	3	
(practice)		5	2	24
		7	4	
	B₃ (5/wk)	8	5	
		7	6	39
		9	4	
				Grand sum
Sums of A		43	29	72

Steps in the Analysis

1. Arrange the data into cells by factors. In this example, factor A is athletic experience (A_1 = experience, A_2 = no experience). Factor B is days of practice (B_1 = one day/week, B_2 = three days/week, and B_3 = five days/week).

2. Determine the cell means and the marginal mean values (averages across factors) for A_1, A_2, B_1, B_2, and B_3. See table 11.2.

Table 11.2 Cell Means and Marginal Means

| | | Factor A (experience) | | Marginals |
		A₁ (athletes)	A₂ (nonathletes)	for B
	B₁ (1/wk)	1.33	1.67	1.50
Factor B	B₂ (3/wk)	5.00	3.00	4.00
(practice)				
	B₃ (5/wk)	8.00	5.00	6.50
Marginals for A		4.78	3.22	4.00
				Grand mean

3. Notice that the marginal values represent the mean values down each column of factor A or across each row of factor B. In other words, the average for all experienced athletes (A_1) regardless of the practice schedule in which they were placed is 4.78. Likewise, the average for nonathletes (A_2) is 3.22.

To determine if athletes benefited from practice more than nonathletes, compute an F value (simple analysis of variance) between these two marginal mean values. This is called the main effect of factor A.

4. Note that the average for all subjects (athletes and nonathletes combined) in the 1-day group (B_1) is 1.5, likewise, for the 3-day group (B_2), the mean is 4.00, and for the 5-day group (B_3) it is 6.5.

To determine if days of practice had an effect on learning, compute an F value (simple analysis of variance) on these three marginal means values. This is called the main effect of factor B.

5. Using a computer, compute the F value for interaction. Remember that interaction is the combined effect of the factors A and B on the dependent variable. If athletes learn at the same rate as nonathletes, then the lines in figure 11.7 will be parallel; that is, they will have the same or nearly the same slope. If athletes benefit more from practice than nonathletes, then the slope of the line for athletes will be greater (steeper) than the line for nonathletes. The interaction analysis computes an analysis of variance between the slopes of the lines.

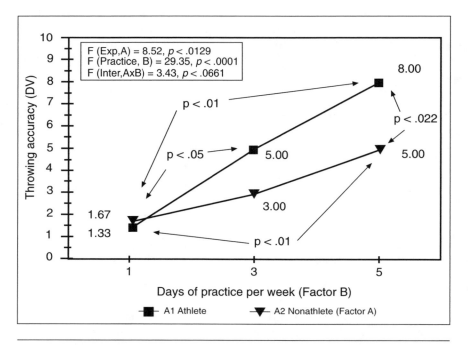

Figure 11.7   Factorial between-between.

Interaction may be the most important finding in the study. Interaction can take several forms. Figures 11.1 to 11.6 represent a design like the one we have been studying—the effects of two levels of A, and three levels of B on a dependent variable.

When interaction is not significant (figures 11.1, 11.2 and 11.3), it may be appropriate to conduct further analyses on the main effects by using simple ANOVA at each level of A and B followed by post hoc tests. If interaction is significant and main effects are significant but ordinal (the mean values for A_1 and A_2 are in the same order at all levels of B), as is exemplified by figure 11.4, perform further analysis on the main effects by using simple ANOVA and post hoc tests if appropriate.

When interaction is significant and disordinal, as in figures 11.5 and 11.6, main effects may be clouded because the average of A_1 and A_2 will be similar across all levels of B. The average of B_1, B_2, and B_3 may also be similar across all levels of A. This can result in a nonsignificant main effect even though the mean values for the cells are considerably different. Under these conditions, a careful review of the data and the potential causes of disordinal interaction effects as they relate to the purpose of the study is needed before a decision to further analyze main effects is made.

Table 11.3   Factorial ANOVA

Source	SS	df	MS	F	p
Factor A (experience)	10.89	1	10.89	8.52	.0129
Factor B (practice)	75.00	2	37.50	29.35	.0000
A × B (interaction)	8.78	2	4.39	3.43	.0661*
Error	15.33	12	1.28		

* Because the p value for interaction is greater than .05, the researcher must decide if a value of .0661 will be considered significant. This determination must be made based on the consequences of making a type I error. For the purposes of this discussion, $p = .0661$ will be considered to be significant.

When these three F values (F for factor A, F for factor B, and F for interaction) have been calculated, a factorial ANOVA table (see table 11.3) is prepared to summarize the results.

Interpreting the Results

The F values for factors A and B represent an ANOVA on the marginal means. These results are called the **main effects.** The significant F for factor A (athletic experience) indicates that for all three levels of B combined, athletes (A_1 marginal mean = 4.78) scored significantly better than nonathletes (A_2 marginal mean = 3.22) (See table 11.2 and figure 11.7). The significant F for factor B (practice) indicates that for both levels of A combined, there is a significant difference somewhere among the three levels of B (B_1 marginal mean = 1.50, B_2 marginal mean = 4.00, and B_3 marginal mean = 6.50). The F for practice does not indicate which of the three levels of B differ from each other. A post hoc test is needed to identify the location of significant cell mean differences.

The interpretation of a factorial ANOVA is a **step down process**. First, the main effects and the interaction are evaluated. If these F values are not significant, the analysis stops here. If any of them are significant, further analysis is warranted. The purpose of the step down procedure is to protect against type I errors. The familywise error rate (see chapter 9) would be increased considerably if all possible cell mean differences were analyzed without first checking for main effects on each factor.

When none of the main effects or interaction is significant, to proceed further (someone called it data snooping) is to risk making a type I error by encountering the difference that occurs 1 out of 20 times at $p = .05$ or 1 out of 100 times at $p = .01$ by chance alone. When main effects or interaction are significant, however, further analysis is appropriate because we are confident that additional differences exist. It is now just a matter of finding them.

To find them we conduct a simple ANOVA across A at each of the levels of B, and across B at each of the levels of A. The results of a step down analysis on the problem under consideration are presented in table 11.4

Simple ANOVAs across A at B_1, B_2, and B_3 reveal that the only significant difference between athletes (A_1) and nonathletes (A_2) occurs at 5 days of practice (B_3),

Table 11.4   Step Down Analysis of Factorial ANOVA

Main Effects

$F_A = $ 8.51, $p < .0129$

$F_B = 29.30$, $p < .0000$

$F_{AB} = 3.43$, $p < .0661$

Simple Effects (A)

F across A at $B_1 = $ .50, N.S.

F across A at $B_2 = $ 2.40, N.S.

F across A at $B_3 = 13.47$, $p < .022$

Simple Effects (B)

F across B at $A_1 = 18.81$, $p < .003$, $HSD^* = 3.34$, $p < .05$; $4.87, p < .01$

F across B at $A_2 = 10.86$, $p < .011$, $HSD^* = 2.21$, $p < .05$; $3.22, p < .01$

*HSD = Tukey's honestly significant difference

(F = 13.47, $p < .022$). Since there are only two groups in this ANOVA (A_1 and A_2), no further analysis with a post hoc test is necessary. Notice that a t test would be sufficient here because only two mean values are being compared. Since $t^2 = F$ when $k = 2$, both t and F will produce the same determination of significance in this case.

The results of the athletes vs. nonathletes throwing accuracy study are graphically presented in figure 11.7.

The **simple effects** across A indicate that the consequence of classification as athlete or nonathlete is not remarkable until 5 days of practice are conducted. Classification by athletic experience does not produce significant differences at 1 or 3 days of practice. The simple effects for A also reveal that the interaction effect (while beginning to take effect at 3 days of practice) only becomes significant at 5 days of practice.

A second set of simple effects tests are performed across B at A_1 and A_2. For B_1 vs. B_2 vs. B_3 at A_1 (athletes), $F = 18.81$, $p < .003$, and Tukey's $HSD = 3.34$ at $p < .05$, and 4.87 at $p < .01$. Applying this information to figure 11.7, it can be observed that the athletes who practiced 3 days are significantly better at $p < .05$ than are the athletes who only practiced 1 day $(5.00 - 1.33 = 3.67)$. Because the interaction is ordinal, it follows that the 1-day-per-week athletes also differ significantly from the 5-days-per-week athletes $(p < .01)$, but 3 days per week does not differ from 5 days per week sufficiently to exceed HSD $(8.00 - 5.00 = 3.00$ does not meet the minimal requirement for HSD at $p < .05$ of 3.34).

The F value for simple ANOVA across B_1, B_2, and B_3 at A_2 (nonathletes) is 10.86, $p < .011$, and $HSD = 2.21$ at $p < .05$ and 3.22 at $p < .01$. Figure 11.7 indicates that for nonathletes, it takes 5 days of practice to produce a significant effect on throwing accuracy $(5.00 - 1.67 = 3.33)$; 3 days is not enough $(3.00 - 1.67 = 1.33)$. In addition, 3 days of practice does not differ from 5 days.

The significant F for interaction (A × B) indicates that athletes responded to the practice differently than did the nonathletes. Figure 11.7 reveals that the lines for athletes and nonathletes are diverging. This divergence represents a difference in the slope of the lines. At 3 days of practice, the lines are not sufficiently divergent to declare the difference significant, but by 5 days, the differences between the lines are significant. This confirms that the interaction has taken place between the 3-day and the 5-day condition, but not between the 1-day and the 3-day condition.

The Magnitude of the Treatment (Size of Effect)

The proportion of the total variance that may be attributed to each factor and interaction can be computed with a redefinition of the formula for omega squared (see equation 9.17). The values for SS, MS, and df may be obtained from the factorial ANOVA table 11.3. For each factor, omega squared is:

$$\omega^2_{fac} = \frac{SS_{fac} - df_{fac}(MS_E)}{SS_T + MS_E}$$

(11.01)

and for interaction:

$$\omega_{AB}^2 = \frac{SS_{AB} - (k_A - 1)(k_B - 1)(MS_E)}{SS_T + MS_E} \qquad (11.02)$$

For factor A (athletic experience):

$$\omega_A^2 = \frac{10.89 - 1(1.28)}{110.0 + 1.28} = .09$$

For factor B (practice frequency):

$$\omega_B^2 = \frac{75.00 - 2(1.28)}{110.0 + 1.28} = .65$$

For interaction:

$$\omega_{AB}^2 = \frac{8.78 - (2 - 1)(3 - 1)(1.28)}{110.0 + 1.28} = .06$$

These calculations indicate that practice is by far the most important factor contributing to the differences in the scores. Practice contributes 65% of the total variance, athletic experience contributes only 9%, and the combined effect of experience and practice adds another 6%. The remainder of the variance is unexplained (error). See Keppel (1991, p. 221-224) for further discussion of the process for evaluating of the magnitude of the treatments.

Conclusions

Factorial ANOVA is a method of simultaneously determining the effects of two or more factors on a dependent variable. In our hypothetical study, the effects of prior athletic experience on a collegiate-level basketball team (factor A) were compared with the effects of one day, three days, or five days per week of practice (factor B) on accuracy in a novel throwing task (the dependent variable). It was determined that the athletic experience did have an effect on the subject's ability to profit from practice, but the difference is only significant if five days of practice per week are conducted ($p < .022$). One day or three days are not sufficient.

The amount of practice was also significant. For the athletes, three days ($p < .05$) or five days ($p < .01$) are better than one day. For the nonathletes, it takes five days of practice per week to observe a significant difference ($p < .01$) in throwing accuracy. Three days per week are no different from one day per week.

Finally, the interaction was significant. Athletes may make more effective use of practice than do nonathletes. Given the same conditions of practice, athletes benefit more from additional days of practice than do nonathletes. The lines depicting the ability of athletes and nonathletes to benefit from practice are not parallel. The line for athletes has a greater slope than does the line for nonathletes.

The relative importance of each factor and interaction was determined with omega squared. Practice was found to be the major factor affecting the dependent variable, contributing 65% of the total variance in the throwing scores. Athletic experience and interaction contributed 9% and 6%, respectively.

You may wonder why there is no difference between athletes and nonathletes at three days of practice (figure 11.7). The lines are diverging at three days, but the difference of two points is not significant $(5.00 - 3.00 = 2.00)$. This may be the result of insufficient power in the study. A power analysis to determine the minimum N needed per cell to have an 80% chance of detecting real differences as small as two points should have been performed before the study began. Perhaps with a larger N, the difference between A_1 and A_2 at B_2 would be significant. See chapter 8 for a discussion of power.

A Between-Within Example

Factorial ANOVA with repeated measures on one factor is often referred to as a **between-within,** or **mixed model** design. "Between" signifies that the analysis on factor A is an ANOVA between independent groups. Each of the groups in factor A consists of different subjects randomly assigned to the groups. "Within" signifies that the analysis on factor B is a repeated measures analysis. The subjects in each of the groups are measured two or more times.

The mixed model is a very common design in kinesiology and many other physical and behavioral sciences. Exercise physiologists often want to know the effects of various training schedules on physiological variables during the training program. Subjects may be placed into treatment groups (high, medium, and low intensity for example) and then trained for a period of weeks. Measurements of VO_2max, heart rate max, respiratory exchange ratio (RER), and other physiological variables are measured weekly throughout the training period to determine the effects of treatment over time.

Motor behavior researchers also use this design to evaluate the effects of various learning schedules on speed and amount of learning over multiple trials or over an extended period of time. This design calls for two or more groups of subjects to be tested two or more times. Hence, the design is a combination of a simple ANOVA (between) and a repeated measures ANOVA (within). Both ANOVAs are conducted simultaneously.

As in the between-between example presented earlier, this analysis also produces three F values. One for the between analysis (groups), one for the repeated measures analysis (trials), and one for interaction (groups $\times$ trials). Interaction is interpreted in this design in the same manner that it was interpreted in the factorial ANOVA (between-between) design. Often, interaction is the most interesting and important finding in a between-within design.

You will recall that interaction identifies the combined effects of group assignment and repeated measurement. Frequently, groups will respond to treatment in a different manner over time. High-intensity training may produce a more rapid improvement in physiological variables than does moderate- or low-intensity training. Certain teaching methods may produce more rapid learning of a specific skill than do other methods. Because of its utility, this design is commonly reported in professional literature.

The *assumptions* for a between-within design are the same as for the between analysis in a simple ANOVA (chapter 9), and the within analysis in ANOVA repeated measures (chapter 10). However, the sphericity assumption (discussed in chapter 10) for the repeated measures factor must now be applied to the pooled data (across all of the groups) as well as to each individual group. This pooled condition is referred to as multisample sphericity, or **circularity** (Schutz and Gessaroli, 1987). Advanced computer programs test for this assumption and provide the epsilon values for the Greenhouse-Geisser and Huynh-Feldt corrections.

To demonstrate this design, another study with simple, hypothetical data will be described. When the concepts from this simple study are understood, more complicated studies with unequal *N*s and a different number of levels of each factor may be considered. Analysis of this design is almost never computed by hand. The conceptual analysis is presented here (without detailed equations) to guide you through the process. An analysis of this nature should always be done by computer to reduce the possibility of calculation error and to save time and effort.

A researcher wanted to investigate the relative effects of distributed and massed practice (factor A—between groups) on the learning over time (factor B—within trials) on a novel motor skill as measured by time on target on the pursuit rotor apparatus (dependent variable). Fifteen college-age students were randomly placed into three practice groups with five subjects in each group. The groups were then randomly assigned to one of the following three practice conditions for the 10-day study.

1. Distributed practice: the subjects practiced in 30-second trials with a 30-second break between trials for a total of 10 trials (5 minutes total practice time each day for 10 days).
2. Massed practice: the subjects practiced for 5 minutes per day without a break for 10 days.
3. Control: the subjects took the pre-, mid-, and posttests, but practiced on an unrelated gross motor task for 5 minutes each day for 10 days.

The dependent variable for the pre-, mid-, and posttests was time on target during the pursuit rotor task for a 15-second period. The dependent variable was measured prior to the 1st practice day (pre), after the 5th day (mid), and after the 10th day (post). This design with three groups measured over three trials would be categorized as a 3 × 3 between-within factorial ANOVA.

The factorial ANOVA will produce an *F* value for the differences *between* groups over the pre-, mid-, and posttests, and an *F* value for the differences *within* the three tests (trials) for each group. An *F* value for interaction, which will determine the combined effects of the practice conditions and the three trials, will also be calculated. The interaction will compare the relative speed of learning of the three groups.

Steps in the Analysis

1. Arrange the raw data (*X*) into a matrix with factor A (groups) on the vertical axis and factor B (trials) on the horizontal axis. Compute the sum of each group and for each trial and the grand sums (see table 11.5).
2. Set up a matrix for the mean values of factor A crossed with factor B (table 11.6.).
3. Using a computer, compute the between *F* value across rows on the marginal means for factor A (groups). This is called main effects for factor A.
4. Using a computer, compute the within *F* value across columns on the marginal means for factor B (trials). This is called main effects for factor B.
5. Using a computer, compute the *F* value to determine interaction among groups and trials.
6. Create a factorial ANOVA table to summarize the results (table 11.7).

Table 11.5   Raw Data and Factor Sums

	B_1 (Pre)	B_2 (Mid)	B_3 (Post)	A Sum
	2	5	10	17
	3	7	11	21
A_1 (Dist.)	5	9	13	27
	4	8	9	21
	3	6	12	21
B Sum	17	35	55	107
	2	7	9	18
	4	5	8	17
A_2 (Mass)	3	3	7	13
	3	5	8	16
	2	8	7	17
B Sum	14	28	39	81
	2	4	6	12
	3	3	5	11
A_3 (Cont.)	2	5	6	13
	3	2	8	13
	3	3	4	10
B Sum	13	17	29	59
Grand sum	44	80	123	247

Factor A = groups (between).
Factor B = trials (within).

Table 11.6   Cell Means

	B_1 (Pre)	B_2 (Mid)	B_3 (Post)	Marginals A
A_1 (Dist.)	3.4	7.0	11.0	7.13
A_2 (Mass)	2.8	5.6	7.8	5.40
A_3 (Cont.)	2.6	3.4	5.8	3.93
Marginals B	2.93	5.33	8.20	5.49 Grand mean

Factor A = groups (between).
Factor B = trials (within).

Table 11.7   Factorial ANOVA, Between-Within

Source	SS	df	MS	F	p
Factor A (Groups)	76.98	2	38.49	19.03	.0002
Error A	24.27	12	2.02		
Factor B (Trials)	208.58	2	104.29	67.77	.0000
A × B (Interaction)	26.49	4	6.62	4.30	.0091
Error B	36.93	24	1.54		

Epsilon: GG = .9088, HF = 1.000.

Adjustment for trials

GG adj. values: F = 67.77, df = 1.82,   21.81, p = .0000.
HF adj. values: F = 67.77, df = 2.00,   24.00, p = .0000.

Adjustment for interaction

GG adj. values: F = 4.30,   df = 3.64,   21.81, p = .0118.
HF adj. values: F = 4.30,   df = 4.00,   24.00, p = .0091.

The F value for main effects on factor A (19.03) is a simple ANOVA on the marginal means (7.13, 5.40, and 3.93) for factor A (groups) (see table 11.6). This F value indicates that there is a difference somewhere among the scores of all subjects in the three groups averaged over all three trials. Further analysis with a simple

ANOVA between the three groups at each trial is justified. If any of these simple effects are significant, a post hoc test to determine cell mean differences is performed. If the F value for main effects for A is not significant, the analysis for factor A stops here. Because this is not a repeated measures analysis, we do not have to worry about the assumption of circularity.

The F value for main effects on factor B (67.77) is a repeated measures ANOVA on the marginal means (2.93, 5.33, and 8.20) for factor B (see table 11.6). This is called the main effects for B. This F indicates that there is a difference somewhere among the mean scores of the three trials averaged over all three groups. Further analysis with a repeated measures ANOVA within each group over the three trials is justified. If any of these simple effects is found to be significant, a post hoc test to determine the cell differences is performed. If the F value for main effects for B is not significant, the analysis for factor B stops here. Since this is a repeated measures analysis, we must check for a possible violation of the assumption of circularity by determining the size of the epsilon value.

Epsilon values (GG = .9088, conservative, and HF = 1.000, liberal) confirm that the violation was minimal. Remember that an epsilon value of 1.00 indicates no violation, and a value > .75 indicates minimal violation. Huynh-Feldt indicates no violation, and Greenhouse-Geisser indicates very little violation. The GG and HF adjustments do not change the p values for the trials analysis. This indicates that the small violation that occurred is not remarkable.

The significant F value for interaction (4.30) reveals that the three groups responded differently over the three trials. Figure 11.8 suggests that the distributed practice group improved their scores over the three trials faster than either the massed or control group. In other words, the lines on the graph of groups by trials are not parallel. Simple ANOVA across groups at the pre-, mid-, and posttests reveals that the differences are significant at the mid- and posttests (table 11.8). After Greenhouse-Geisser or Huynh-Feldt adjustments for violation of circularity (table 11.7), p values are still highly significant (HF indicates no change at all since HF epsilon = 1.000) corroborating that there was minimal, if any, violation of the assumption.

Note: If the GG or HF adjustments reduce the p values below the preset alpha level for rejection of the null, the data may be reanalyzed using MANOVA (chapter 12). In this case, the repeated measures are treated as multiple dependent variables and the assumptions of sphericity and circularity are not required. If the F value for trials and interaction is still significant under MANOVA analysis (which is less powerful than the repeated measures design), the null hypothesis may be rejected with the confidence indicated by the MANOVA analysis. See Schutz and Gessaroli (1987) for further discussion of this issue.

7. Conduct a step down analysis from main effects to simple effects to cell differences and present in table form (table 11.8).

The ANOVAs for simple effects and post hoc tests are conducted in the manner delineated for a single factor ANOVA described in chapter 9 (independent groups) and chapter 10 (repeated measures). You will notice that the values for Tukey's honestly significant difference test for determination of cell differences are the same

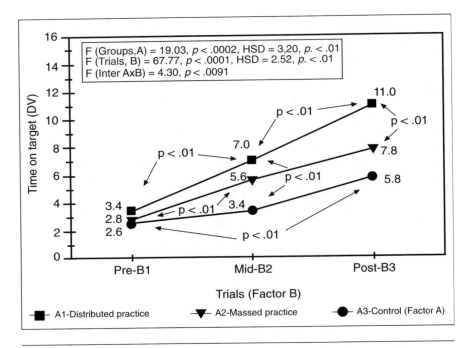

Figure 11.8   Factorial between-within.

Table 11.8   Step Down Analysis of Factorial ANOVA

Main Effects

$F_A = 19.03$, $p < .0002.$

$F_B = 67.77$, $p < .0000.$

$F_{AB} = 4.30$, $p < .0091.$

Simple Effects (A)

F across A at $B_1 = $ 1.13, N.S.

F across A at $B_2 = $ 6.50, $p < .0122$ *HSD = 3.59, $p < .01.$

F across A at $B_3 = $ 19.11, $p < .0002$ HSD = 3.02, $p < .01.$

Simple Effects (B)

F across B at $A_1 = $ 46.96, $p < .0000$ HSD = 2.52, $p < .01.$

F across B at $A_2 = $ 20.40, $p < .0000$ HSD = 2.52, $p < .01.$

F across B at $A_3 = $ 9.01, $p < .0012$ HSD = 2.52, $p < .01.$

HSD = Tukey's honestly significant difference

for all simple effects on factor B (repeated measures). This is because the pooled mean square error term for main effects was used to calculate *HSD* in each case as is done in both the BMDP and SPSS software packages. Some difference of opinion exists among statisticians as to whether the main effects error term should be used or if a separate error term for each simple effect should be calculated. There are arguments on both sides of this issue. The interested reader is referred to Tabachnick and Fidell (1989, pp. 45 and 465) for further discussion of this matter.

The *HSD* values could be calculated for $p < .05$ if this level had been set prior to data collection as an acceptable rejection level.

The Magnitude of the Treatment (Size of Effect)

Applying equations 11.01 and 11.02 for omega squared to the data in table 11.7 results in a determination of the relative importance of the factors of groups and trials. For groups:

$$\omega_G^2 = \frac{76.98 - 2\,(2.02)}{373.24 + 2.02} = .19$$

For trials:

$$\omega_T^2 = \frac{208.58 - 2\,(1.54)}{373.24 + 1.54} = .55$$

For interaction:

$$\omega_{GT}^2 = \frac{26.48 - (3-1)(3-1)(1.54)}{373.24 + 1.54} = .05$$

Conclusions

When table 11.4, figure 11.8, and omega squared are interpreted based on the main effects, simple effects, and post hoc tests, the following conclusions seem justified.

1. There is no difference among the groups on the pretest. This confirms that the distribution of subjects into groups by random assignment was effective.
2. There is a significant difference in the amount of learning among the three groups. On the midtest, the distributed group is significantly better than the control group (7.0 –3.4 = 3.6). On the posttest, the distributed group is significantly better than the massed or control groups (11.0 – 7.8 = 3.2) but the massed group is not significantly better than the control group (7.8 – 5.8 = 2.0). Group assignment accounts for 19% of the variance in time on target.
3. The distributed group improved their scores significantly between the pre- and the midtest (7.0 – 3.4 = 3.6) and between the mid- and the posttest (11.0 – 7.0 = 4.0). The massed group improved their scores between the

pre- and the midtests (5.6 − 2.8 = 2.8) but did not continue this improvement between the mid and the post test (7.8 − 5.6 = 2.2). The control group improved their scores between the pre- and the posttest (5.8 − 2.6 = 3.2), but not from pre- to midtest (3.4 − 2.6 = .8) or mid- to posttest (5.8 − 3.4 = 2.4). Since they did not practice the skill, this improvement probably represents learning that took place during the testing sessions. Factor B, trials, is the most important factor to affect the dependent variable. It accounts for more than one-half (55%) of the variance in time on target scores.

4. The groups did not learn at the same rate. Significant interaction indicates that the lines are not parallel. The distributed group learned at a rate faster than the control or massed groups because it differs ($p < .01$) from the control at the midtest, and it differs from both other groups at the posttest ($p < .01$). Since the massed group is not significantly different from the control group at either the mid- or the posttest, we cannot conclude that they were learning faster than the control group. Interaction took place only between the distributed and the other two groups and it accounts only for 5% of the total variance.

A Within-Within Example

A researcher wanted to know if there were any physiological differences in response to exercise on a traditional treadmill compared to a stair-step treadmill. Thirteen healthy male college students were randomly selected and asked to report to the lab on two different days with at least one rest day in-between. On the first day, subjects were asked to perform a graded submaximal exercise test on a treadmill. The test consisted of four increasing stages of work. Physiological responses were recorded at each stage as dependent variables. The amount of physical work performed on the treadmill at each stage was calculated.

On the second day, the same subjects performed an identical amount of physical work over four increasing stages on a stair-step treadmill device. Hence, the same subjects performed on two different modes of exercise (factor A—within) and over four equivalent stages of work (factor B—within). This design may be categorized as a 2 × 4 within-within factorial ANOVA.

The analysis of this design is similar to the analysis of between-within, but since both factors are repeated measures on the same subjects, the assumptions of circularity must be met for both factors. The analysis produces three F values, one for mode of exercise, one for stages of exercise, and one for interaction. A similar step down analysis is performed when either the F for mode or the F for stages is significant. A significant interaction may be the finding of most interest in this study. If interaction is not significant, then subjects respond in the same manner on both modes of exercise to the identical graded exercise stimulus. If the interaction is significant, then physiological responses are different on the two modes of exercise. The mean values for heart rate are presented in table 11.9 and the factorial ANOVA is presented in table 11.10.

Table 11.9   Cell Means and Marginal Means (DV = Heart Rate)

		Stages (Factor B)				
		One (B1)	Two (B2)	Three (B3)	Four (B4)	Marginals (A)
Mode (factor A)	A1-Tread	98	117	133	143	122.25
	A2-Step	99	113	131	142	121.25
	Marginals (B)	98.5	115.0	132.0	142.5	121.75 Grand mean

Table 11.10   Factorial ANOVA, Within-Within

Source	SS	df	MS	F	p
Factor A (Mode)	51.24	1	51.24	.34	.5698
Error A	1800.63	12	150.05		
Factor B (Stages)	29403.26	3	9801.09	177.81	.0000
Error B	1984.37	36	55.12		
A × B (Interaction)	94.72	3	31.57	1.23	.3129
Error A × B	923.90	36	25.66		

Check for circularity:

Mode
Because there are only two values for mode, homogeneity of covariance cannot be violated. Epsilon values are not produced, therefore no adjustment is made.

Stages
Epsilon: GG = .7804, HF = .9813.

Adjustment for stages
GG adj. values: $F = 177.81$, $df = 2.34, 28.09$, $p = .0000$.

HF adj. values: $F = 177.81$, $df = 2.94, 35.33$, $p = .0000$.

Interaction
Epsilon: GG = .6775, HF = .8168.

Adjustment for interaction
Because unadjusted F is not significant, no adjustments are needed.

Step Down Analysis

F for mode is not significant, therefore no further analysis is justified. The *F* value for stages is significant and may be followed by a post hoc test to determine which stages differ. Tukey's *HSD* at $p < .01$ is 9.88. Since every stage for both modes is more than 9.88 beats/minute higher than the previous stage, we can conclude that heart rate increases significantly at every stage. The *F* value for interaction is not significant, indicating no differences in the slopes of the lines. Figure 11.9 presents the data in graphic format.

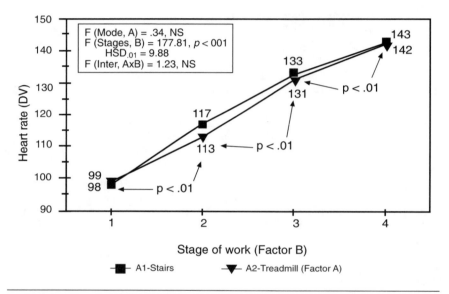

Figure 11.9   Factorial within-within.

Conclusions

1. The *F* value for mode is not significant, therefore we concluded that the heart rate did not differ on the two exercise modes.
2. The *F* value for stages is highly significant. While there was a small violation of circularity for stages, it was not sufficient to alter the *p* value. The increase in heart rate was expected because physical work increased stage by stage. Indeed, if we did not see a significant increase in heart rate over stages, we would suspect an error. Tukey's *HSD* indicates that there was a statistically significant increase in heart rate within each of the four stages.

3. The F value for interaction is not significant. This confirms the insignificant F value for mode and provides further evidence that the subjects responded physiologically in the same manner on both modes of exercise over all four stages.

Summary

Factorial ANOVA is a valuable tool for evaluating the simultaneous effects of more than one factor on a dependent variable. It is commonly used in kinesiology and related disciplines. Theoretically any number of factors can be considered, but typically no more than three are used.

When different subjects are involved in either factor, the design is referred to as a between design. This indicates that the groups are independent. When the same subjects are measured more than once, the design is referred to as a within design. Under these conditions, the assumptions of circularity must be met. Epsilon values to determine the level of violation are calculated by a computer. When circularity is not met, the degrees of freedom are multiplied by epsilon to arrive at an adjusted p value, which will be higher (less significant) than the original p value.

F values for each factor and for the interaction of the factors are calculated. Interaction may be the value of most interest in these studies because it indicates the combined effect of the factors. When the F value for main effects on either factor is significant, further analysis with simple ANOVAs across the factors and post hoc tests may be used to identify differences between and within cell means.

Key Words in This Chapter

Factor	Main effects
Factorial ANOVA	Step down process
Between	Simple effects
Within	Between-within or Mixed model
Interaction	Circularity

Chapter 12

Advanced Statistical Procedures

Most research that compares groups after the application of treatment is based on the assumption that the groups are equal on the dependent variable (DV) prior to treatment. What if they are not? Analysis of covariance (ANCOVA) may be the answer to this dilemma. Researchers often want to analyze more than one dependent variable, yet performing multiple ANOVAs on many DVs may invite a type I error. Multiple analysis of variance (MANOVA) can help solve this problem. What are the factors that make up a complex motor skill? What are the components of physical fitness? Factor analysis can identify the factors that are inherent in a general concept, such as skill or fitness. How can we classify subjects into groups based on common characteristics of the subjects? Discriminant analysis will allow us to do this. These advanced statistical techniques will be discussed in this chapter from an introductory basis only.

Analysis of Covariance

A general assumption of most research designs that use two or more groups treated over time (pre-post design) is that the mean values of the dependent variable for all groups are not significantly different on the pretest. If they are different, and the relative value of the pretest mean difference carries over to the posttest, the posttest values will be confounded by the effects of the pretest differences. Depending on how the pretest group differences are arranged, the posttest means may be farther apart or closer together than would be expected due to treatment alone. This may result in either a type I or type II error.

The recommended way to equate groups on the pretest is to randomly select them from the population being studied. If random assignment is followed and if the number of subjects in each group is reasonably large, it can be assumed that the mean values on the pretest will be equal or nearly so. This can be checked with an ANOVA prior to treatment.

A second method of equating groups prior to treatment is to match them into groups according to their pretest scores as described in chapter 8. The subjects are arranged in order of their pretest scores and placed into groups by the ABBA method. Subject 1 is assigned to group A, subject 2 and subject 3 to group B, subject 4 to group A, and so forth. This assures that the higher of two subjects who are adjacent in the ordered list will not always be assigned to group A. This same method may be used to match into three, four, or any number of groups.

Occasionally, however, the researcher may not be able to randomly assign subjects to groups and may not be able to match the groups. This occurs when subjects are already arranged into groups (i.e., school classes) and the entire group must remain intact in the study. If the pretest group means are not equal, a statistical adjustment in the posttest means must be made to account for the pretest differences. Without this adjustment, the effects of treatment will not be clear.

Analysis of covariance (ANCOVA) is a convenient and useful method of accounting for differences in pretest means or for statistically equating groups on other factors (covariates) that the investigator suspects may influence the dependent variable. When we must contend with groups that differ on covariates and no other way of equating groups is available, ANCOVA is appropriate.

ANCOVA is based on the assumption that the covariate is related to the dependent variable (i.e., they covary). That is, when the scores on the covariate increase or decrease, the dependent variable scores also increase or decrease. The relationship between the covariate and the dependent variable may be expressed by a correlation coefficient. ANCOVA uses regression to adjust the values of the dependent variable to account for differences that may exist among the groups being studied prior to treatment.

It is common to use pretest means as a covariate. However, other variables that affect the dependent variable may also serve as covariates. Such confounding variables as IQ, GPA, strength, aerobic condition, height, weight, percent body fat, or others may serve as covariates if we suspect that they may have a confounding effect on the groups in the experiment. It is possible to have more than one covariate. In this case, the dependent variable is adjusted by multiple regression to simultaneously account for the effects of more than one covariate.

When the pretest group means differ significantly, the pretest becomes a covariate along with the treatment (independent variable) effects. To account for pretest differences ANCOVA is applied to the dependent variable (posttest) scores. Without ANCOVA, the posttest scores vary according to two factors, (1) the covariate effects and (2) the treatment effects. ANCOVA is essentially an ANOVA design with regression used as a method of adjusting posttest scores based on the effects of the covariate. The purpose of ANCOVA is to factor out the effects of the covariate so that the adjusted posttest scores reflect only the effects of treatment.

In the pre-post example, the relationship is obvious. Subjects take the test twice (pre and post) and there is a relationship between their scores. It is this relationship that permits the adjustment of posttest scores based on the covariate using bivariate regression (see chapter 7). When a linear relationship exists, bivariate regression may be used to predict what the posttest means would be if all subjects started at the same point, (the grand mean of the pretest). In other words, the posttest group means are adjusted up or down depending on whether the pretest group means were above or below the pretest grand mean.

This same procedure can be used to adjust the dependent variable based on any covariate that may not be randomly distributed across the groups. For example, if we were comparing the ability of several groups to learn a complicated motor skill under differing treatment conditions, intelligence may be a factor that partially determines a subject's ability to learn the task. In this case, it would be important to know that the mean IQs of each group are equal. If one group contained people with superior intelligence, the subjects in that group may learn at a faster rate because of intelligence, rather than treatment effects. Thus, IQ would be a covariate, or confounding variable.

Ideally these confounding variables should be randomly distributed throughout the groups so that their effects would not influence one group more than another. But when confounding variables are not randomly distributed among the groups, it may be necessary to correct for their influence with ANCOVA. The influence of one covariate may be adjusted using bivariate regression, but multiple regression is needed to calculate the simultaneous influence of two or more covariates.

Assumptions and Cautions

Because the posttest values are adjusted through regression techniques, the amount and the accuracy of the posttest mean adjustment is dependent on the magnitude of the correlation coefficient between the covariate and the posttest values. Unless the correlation is perfect (± 1.00), the adjusted posttest values will always contain some error. For this reason, ANCOVA should not be used unless there is no other way to design the study. Random assignment to groups or groups matched on the covariate are the preferred solutions. ANCOVA should only be used as a last resort, when random assignment or matching is not possible.

The assumptions in ANCOVA (Keppel, 1991, p. 316) are the same as in ANOVA, with the following additional assumptions:

1. A linear relationship between the covariate and the dependent variable. If the relationship is not linear, the adjustment will be minimal and of little or no benefit to the analysis.

2. Homogeneity of the regression coefficients. The slopes of the linear correlations between the covariate and the dependent variable are equal or nearly so across the groups. If this last assumption is violated, ANCOVA may lead to a type II error (i.e., the adjusted F value may be reduced in size). Tests for the assumption of homogeneity of regression coefficients may be found in more advanced textbooks.

Multiple Analysis of Variance

In the t test, simple ANOVA, factorial ANOVA, and ANCOVA, only one dependent variable was considered. For example, the effect of dietary fat levels (independent variable) on body fat (dependent variable) may be studied. This design could be analyzed with a t test (comparing two groups), with simple ANOVA (comparing three or more groups), with a 3×2 factorial ANOVA (comparing three or more groups with an additional factor of activity level, e.g., active vs. inactive), or ANCOVA (comparing two or more groups whose mean values are adjusted to account for the effects of a covariate, such as blood cholesterol levels).

Each of these research designs may be expanded to determine the simultaneous effect of treatment on more than one dependent variable. For example, the effects of dietary fat levels on fat mass, lean body mass, and aerobic capacity may be of interest. In this case, three dependent variables are analyzed simultaneously, hence the term **multiple analysis of variance,** or MANOVA.

MANOVA is useful for several reasons:

1. It helps to protect against type I errors. In MANOVA, a new dependent variable (DV) is formed from the linear composite of the several dependent variables using regression techniques. This new virtual dependent variable is then analyzed with ANOVA to determine if there are any differences among the treatment groups. This is sometimes referred to as the omnibus F value. If omnibus F is not significant, the analysis stops.

If differences are found on the newly formed DV, further analysis of the original DVs using ANOVA may be justified. If ANOVA is significant for any of the DVs, post hoc tests to identify group mean differences are conducted. This step-by-step process may help to reduce the familywise error rate discussed in chapter 9.

MANOVA adds one more step to the step down process previously described for ANOVA (see chapters 9, 10, and 11). At each step, a nonsignificant result usually terminates the analysis. If omnibus MANOVA is significant, this may indicate that one or more of the several DVs differs among the treatment groups. A univariate ANOVA is then performed on each of the separate DVs to determine which one or ones contain the effect. If univariate ANOVA is not significant for a given DV, the analysis stops for that DV. When univariate ANOVA is significant on one of the DVs, post hoc tests may be conducted to identify individual group differences.

Wagoner (1994), however, cautions against this procedure. If the researcher wishes to analyze all of the dependent variables with multiple ANOVAs, Wagoner suggests a Bonferroni adjustment to the original alpha. He asserts that MANOVA is most useful for determining the underlying factors, which may be represented by two or more of the dependent variables. To identify these factors, he recommends that MANOVA be followed by a descriptive discriminant analysis (DDA). The choice of whether to use ANOVA or MANOVA thus depends on the research question being asked. If the underlying constructs (factors) among the dependent variables are of interest, use MANOVA. If each of the dependent variables is of autonomous interest, use ANOVA with a Bonferroni adjustment.

2. With more than one DV, MANOVA offers a greater chance of determining the effects of treatment. The independent variable may affect one DV but not another. Researchers usually collect data on several DVs while the subjects are being tested. This is much more efficient than testing them several times. With the additional data available from several DVs, more information about the phenomenon being studied may be analyzed with MANOVA. However, indiscriminate use of multiple DVs is not an appropriate technique. Before collecting data, the researcher should plan the DVs to be studied based on the logic of the study and the theory behind it. To simply add multiple DVs into the analysis in hopes of finding something significant is poor quality research and an open invitation to a type I error.

3. Under certain conditions, MANOVA may be more powerful than ANOVA. However, this is usually not the case. MANOVA is frequently less powerful (i.e., more conservative) than ANOVA. When several of the DVs are not significant and one is just barely significant, MANOVA is less powerful. The insignificant DVs may mask the effects of the one significant DV producing an insignificant omnibus F for MANOVA. This may cause you to accept the null hypothesis for all of the DVs when it is false for at least one (a type II error).

While MANOVA seems to be a logical extension of ANOVA, which should be used whenever possible, it may not always be advantageous. Tabachnick and Fidell (1989, p. 372) state it this way:

> There are no free lunches in statistics, just as there are none in life. MANOVA is a substantially more complicated analysis than ANOVA. There are several important assumptions to consider, and there is often some ambiguity in interpretation of the effects of IVs on any single DV. Further, the situations

in which MANOVA is more powerful than ANOVA are quite limited; often MANOVA is considerably less powerful than ANOVA. Thus, our recommendation is to avoid MANOVA except where there is a compelling need to measure several DVs.

Assumptions and Cautions

The assumptions that underlie the t test, and simple repeated measures and factorial ANOVA also apply to MANOVA. As in any design, MANOVA may be generalized to a population only if the subjects in the study have been randomly selected from that population. Other assumptions and cautions listed below must also be observed when MANOVA is used.

Common Sense

The finding of a significant F value for MANOVA (as with any other statistical computation) does not assure causality. Causality is a logical determination based on review of literature, proper research design, control of all extraneous variables, and finally, confirmation by statistical analysis. Beginning researchers tend to accept whatever comes from the computer without serious consideration of what went in. This is especially true of very complicated processes like MANOVA. Before the results of the MANOVA are confirmed, a review of the conditions that prompted the research and the design of the study must be conducted to determine that the results make sense from both a logical and a discipline-based perspective.

Relationships Among Dependent Variables

MANOVA uses multiple regression techniques to create a new dependent variable from a combination of the several DVs in the study. As was discussed in chapter 7, multiple regression is most effective if the variables in the prediction are unrelated to each other. In MANOVA, highly related DVs are not useful because they essentially measure the same variance twice. For the most effective use of MANOVA, select DVs that are all pertinent to the research design, but which measure independent factors of interest. Because there are limitations on the number of DVs that can be used, it is not wise to waste degrees of freedom on two or more related variables, only one of which contributes significantly to the analysis.

Subjects-to-Dependent-Variable Ratio

To maintain an appropriate level of degrees of freedom, there must be more subjects per group than the number of dependent variables. Some statisticians feel that the ratio of subjects per group to DVs should be at least 3 to 1. As the ratio approaches 1 to 1, the power of MANOVA becomes severely limited. In addition, with only 1 or 2 subjects per dependent variable, the assumption of sphericity in repeated measures designs (homogeneity of variance-covariance) is likely to be rejected (Tabachnick and Fidell, 1989, p. 377–378). This is a major concern in MANOVA. When univariate analysis is utilized, one may find studies with 2 or 3 subjects per group. This may be acceptable (if there is sufficient power to conduct the analysis) when only one DV is analyzed. But in MANOVA, with 2 or more DVs, more subjects per group are required.

If you are faced with a low subject-per-group to DV ratio (with no way to measure more subjects or reduce DVs), do not use MANOVA. Instead, do as Wagoner (1994) suggests and perform ANOVA on each DV, then adjust the alpha level with a Bonferroni adjustment based on the number of ANOVAs performed. This is a reasonable solution to the problem of low subject-per-group to DV ratio.

Outliers

Outliers, values that are significantly beyond the range of typical scores in the data set, can seriously affect the results of MANOVA (see chapter 7 for a review of the problem of outliers). Outliers should be identified and corrected or eliminated prior to the analysis. As discussed in chapter 7, outliers may be univariate or multivariate. Multivariate outliers are particularly hard to find in a large data base but have significant effects on MANOVA.

Outliers may result from measurement or recording errors, or they may represent a case that is not typical of the population from which the samples were drawn. If the outliers are the result of measurement or recording errors, they must be found and corrected. If they result from real data points that lie well beyond the other values in the study, you must evaluate each one to determine if it is representative of the subjects to be studied. Failure to remove or correct significant outliers may invalidate the results of a MANOVA analysis.

Repeated Measures Designs

When repeated measures are used with multiple dependent variables, the chances of violating the assumptions of sphericity or circularity are dramatically increased. If the Greenhouse-Geisser or Huynh-Feldt adjustments to the p values do not adequately correct for the violation, an alternate solution using MANOVA in a doubly multivariate design is possible. Doubly multivariate means that the usual DVs serve as one set of multiple variables, and the repeated measures serve as a second set of multiple variables. Under these conditions, the assumptions of sphericity and circularity are not required. See Schutz and Gessaroli (1987) for an in-depth discussion of this procedure.

Interpreting the Results

MANOVA may be applied to any research design where ANOVA is appropriate but where there is more than one DV. The mathematics to compute MANOVA are complicated, and MANOVA is always performed on a computer. The student who understands the basics of ANOVA, repeated measures ANOVA, and factorial ANOVA will be able to apply this knowledge to interpret MANOVA.

Factor Analysis

When data are collected on many variables using the same set of subjects, it is possible, indeed probable, that some of the variables are related to each other. For example, if we wanted to determine the overall motor skill and fitness components of a group of people, we might collect data on the following variables: chin-ups, curl-ups in one minute, bench press, leg press, 50-yard dash, 440-yard dash, one-mile run, sit and reach, skinfolds, softball throw for distance, softball throw for accuracy, standing long jump, vertical jump, hip flexibility with a goniometer, push-ups, step test for aerobic capacity, max VO_2 on a treadmill, body mass index, shuttle run, stork stand for balance, Bass stick test, reaction time, and perhaps many others.

You may immediately say, "Wait, we don't need to use all of those tests because some of them are testing the same component of skill or fitness." What you really mean is that there are correlations between or among many of the measured variables. When two items are highly correlated, we can surmise that they are essentially measuring the same factor. For example, the sit-and-reach test and the hip flexibility with a goniometer test would be expected to be highly related to each other, because the sit-and-reach test measures hip and low-back flexibility, while the goniometer test measures just hip flexibility. The common factor of hip flexibility produces a correlation between the variables. Likewise, bench press, push-ups, and chin-ups are all measures of upper body strength but in slightly different ways. The one-mile run and max VO_2 on a treadmill are also highly related. They both measure factors related to aerobic capacity.

One could discover these relationships by producing an intercorrelation matrix among all the variables and look for r values in excess of some predetermined level, say ± .80. With a few variables, this may work fairly well, but when many variables are measured, the intercorrelation matrix grows rapidly and soon becomes too cumbersome to deal with manually.

To resolve this issue, a statistical technique called **factor analysis,** which will determine the common component of two or more variables and identify that component as a "factor," has been developed. The factor is derived (virtual); that is, it does not actually exist as a stand-alone measured value. It is simply the result of the common component between or among two or more actually measured variables. The factor has no name until the researcher gives it one. For example, the relationship among bench press, push-ups, and chin-ups produces a factor that the researcher might label "upper body strength."

Notice that we have not actually measured upper body strength; we have measured the number of bench presses, push-ups, and chin-ups a person can do. But to determine the "upper body strength" of the subjects, we do not need to measure all three variables. Since the correlation among the variables is high (e.g., $r > .80$), if a person can do many bench presses, we can predict that he or she will also be able to do many chin-ups and push-ups. Hence, we only need to measure one of the variables to get an estimate of the subject's upper body strength. Using this technique, researchers have identified the most common components of physical fitness: muscular strength, muscular endurance, aerobic capacity, body composition, and flexibility.

The same argument could be used for measures of speed. Perhaps we would find a correlation among the 50-yard dash, the 440-yard dash, and the shuttle run. The common factor here is sprinting speed. However, because each variable has unique characteristics (e.g., the shuttle run also requires the ability to stop and change directions quickly), the correlation among them is not perfect. But there is a common factor among them that we could label "sprint speed."

Factor analysis is therefore just a method of reducing a large data set to its common components. It removes the redundancy among the measured variables and produces a set of derived variables called factors. We have seen this before in our study of multiple regression. In figure 7.8 on page 109, variable Y shares common variance with X_1, X_2, X_3, and X_4 but very little with X_5. Variables X_3 and X_5 are highly correlated (colinear) with each other, so they contain a common factor. Variable X_5 is redundant and does not need to be measured, because X_3 can represent it in their common relationship with Y.

Computer programs that perform factor analysis will list the original variables that group together into common derived factors. These factors are then given a number called an **eigenvalue.** The eigenvalue indicates the number of original variables that are associated with that factor (Kachigan, 1986, p. 246). If we are dealing with 10 original variables, then each one contributes an average of 10% of the total variance in the problem. If factor 1 had an eigenvalue of 4.26, then that factor would represent the same amount of variance as 4.26 of the original variables. Since each variable contributes an average of 10%, factor 1 would contribute 4.26×10 or 42.6% of the total variance.

The more variables that load on a given factor (i.e., its eigenvalue), the more important that factor is in determining the total variance of the entire set of data. By comparing eigenvalues, we can determine the relative contribution of each factor. This method will usually identify the two or three most important factors in a data set. Since it is the goal of factor analysis to reduce the number of variables to be measured by grouping them into factors, we are looking for the least number of factors that will adequately explain the data. Generally speaking, when the eigenvalue for a factor drops below the average variance explained by the original variables (in our example, 10%) it is not used. It is rare to identify more than three or four factors.

After the factors are identified, the computer can perform a procedure called "factor rotation." This is a mathematical technique where the factors are theoretically rotated in three dimensional space in an attempt to maximize the correlation among the original variables and the factor on which they load. It also attempts to reduce the relationships among the factors. After all, if two factors are highly correlated to each other, we do not need both of them. Orthogonal rotation is the method where the computer attempts to adjust or rotate the factors that have been selected as most representative of the data (i.e., the ones with the highest eigenvalues) so that the correlation among the factors is orthogonal (as close to zero as possible) and the correlation between the variables that load on a factor and the factor is maximized.

Assumptions and Cautions

Sample size is critical in factor analysis. Tabachnick and Fidell (1989, p. 603) suggest at lease five cases for each measured variable. Sample size is less critical when the factors are strong and distinct. Variables should be reasonably normal; one should check them with a skewness and kurtosis test. Factor analysis assumes that the correlation among the variables is linear. If any correlation is curvilinear, the analysis will be less effective. Transformations of data can sometimes reduce curvilinearity. Multicollinearity or singularity is usually not a problem because we are actually trying to identify these characteristics of the data. If two variables are perfectly related (singular), they are not both needed; one can be eliminated. Finally, outliers must be identified and eliminated or transformed. Outliers will reduce the ability of the computer to find the appropriate factors.

Discriminant Analysis

With the t test and ANOVA, we tested group means to determine if there were significant differences between or among groups. Discriminant analysis is somewhat the opposite procedure. Using regression equations, **discriminant analysis** attempts to classify subjects into groups (the dependent variables) based on certain characteristics of the subjects (the independent variables). In other words, we are attempting to discriminate among the subjects by classifying them into certain groups. For example, we may want to measure subjects on a questionnaire of their life style habits (IVs) in order to classify them as fit or not fit (DVs). To evaluate our group assignments, we could perform a chi-square test to determine if a nonparametric classification is better than chance, or we could perform a t test on an acceptable parametric measure of fitness (say max VO_2) to determine if the group means differ significantly. If they do, this would provide evidence to validate the classification produced by the discriminant analysis.

This procedure is similar to regression analysis and is used for basically the same purpose, but in regression, one attempts to predict the value of a continuous dependent variable; and in discriminant analysis one tries to predict discrete group membership. Fitness consultants could distinguish among a group of people as to whom should receive certain exercise prescriptions, medical researchers could determine who is most likely to be at risk for certain diseases, and employers could determine who among the applicants for a job would be the most promising employee.

Sometimes the independent variables are called predictor variables, and the dependent variables are called criterion variables. The criterion variable must have at least two values, but it may have more. For example, to discriminate more precisely among subjects regarding their fitness levels, one might classify them as highly fit, moderately fit, and unfit.

Discriminant analysis is sometimes incorrectly used when regression analysis should be used. If the DV is dichotomous or has several categories but is not continuous, discriminant analysis is appropriate. However, if the DV is continuous, then regression analysis should be used. If discriminant analysis is inappropriately used on a continuous variable, it will force the dichotomization of the continuous variable (i.e., consider it as nominal data) when it should be analyzed as interval or ratio data.

Assumptions and Cautions

As with other parametric procedures, certain assumptions apply to discriminant analysis. First, it must be assumed that random selection of subjects from the population has taken place and that the distribution of the IVs and DVs in the population is normal. In addition, the variance of the predictor variables (IVs) must be the same in the two or more populations into which the subjects will be classified. Second, the correlation between any two predictor variables must be the same in the two populations into which the subjects will be classified (Kachigan, 1986, p. 219). Finally, it is assumed that the relationships among the IVs and DVs are linear. To improve the results, outliers should be identified and transformed or eliminated prior to analysis, and there should be a check for multicollinearity and singularity among the IVs (Tabachnick and Fidell, 1989, p. 512).

Summary

Because advanced techniques are always performed on a computer, it is not necessary that you understand all of the formulas and mathematical calculations that produce the result. But it is critical that you understand what the answer means, how to interpret it, and the assumptions that underlie it. This chapter has discussed four advanced statistical techniques: ANCOVA, MANOVA, factor analysis, and discriminant analysis. You should be aware of their existence, be able to describe the conditions under which they are applied, and be able to consult statistical software programs and advanced statistical texts to make full use of these procedures.

Key Words in This Chapter

Analysis of covariance
Multiple analysis of variance
Factor analysis

Eigenvalue
Discriminant analysis

Chapter 13

Analysis of Nonparametric Data

A physical education teacher believed that the ability to serve well was related to the success or failure of beginning tennis students. To test this belief, she conducted a ladder tournament with beginning players to determine her students' abilities to win games against their peers. When the tournament was over, she ranked the students according to the order they finished in the tournament. Next, she administered a serving test that ranked the students from best to worst on serving skill. The resulting two sets of rank order scores represent ordinal data that are nonparametric (the assumptions of normality do not apply). How can the teacher analyze these data?

As discussed in chapter 1, data can be classified into four categories: nominal, ordinal, interval, and ratio. When ratio or interval data are collected, analysis by parametric statistical techniques is appropriate. Pearson's correlation coefficient, the *t* test, and ANOVA in all of its varieties are parametric statistical techniques. But when data are of the nominal or ordinal type, the assumptions of normality are not met, and nonparametric statistical techniques must be used.

This chapter presents several of the most commonly used nonparametric statistical techniques:

- **Chi-square** compares two or more sets of nominal data that have been arranged into categories by frequency counts.
- Spearman's **rank order correlation** coefficient determines the relationship between two sets of ordinal data.
- The **Mann-Whitney *U* test** determines the significance of the difference between ordinal rankings of two groups of subjects ranked on the same variable.
- **Kruskal-Wallis ANOVA by ranks** is used to compare the ranking of three or more independent groups. It is similar to simple ANOVA.
- **Friedman's two-way ANOVA by ranks** is similar to repeated measures ANOVA. It determines the significance of the difference between ranks on the same subjects.

Chi-Square (Single Classification)

Chi-square is used to compare two or more sets of nominal data that have been arranged into categories by frequency counts. Gender is an example of nominal data. A person is classified as male or female, and we simply count the number in each category. There is no variability within the category; all subjects are of equal value. Classification must be mutually exclusive; a subject may be classified into only one of the categories.

Grades are another example of this type of classification. After considering all of a student's test scores, the teacher assigns the student a grade for the course. The grade is just a name (nominal data) for performance that has a certain meaning. The grades do represent ordered values (A, B, C, D, and F), but we will treat them here as nominal by counting the number of students in each category.

A certain number of students (the frequency) receive A's, others B's, C's, D's, and F's. When this is completed, all the A's are equally valuable, all the B's equally valuable, and so on for the C's, D's, and F's. The chi-square test should be used to compare two sets of data, classified by category, to determine if the frequencies of the categories differ by amounts larger than would be expected by chance. It reveals the significance of the differences in the frequency counts. The example following shows how chi-square may be applied.

An Example From Administration

A physical education department established a policy that teachers should award certain percentages of the various grades in their classes. After much discussion, the teachers agreed to keep the grade distribution to approximately 10% A's, 30% B's, 35% C's, 20% D's, and 5% F's. The teachers acknowledged that for a small class these percentages might not hold true, but when all classes for a given teacher are combined, the final grade percentages should closely approximate the established policy.

Everyone accepted the policy, and for a year or two the faculty followed it closely. But then one teacher began to consistently give higher grades than would be expected by the policy. In a roster of 141 students, this teacher gave 30 A's, 57 B's, 32 C's, 15 D's, and 7 F's. The chairperson wanted to know whether or not such a distribution differed significantly from the accepted policy. Based on 141 students, departmental policy would call for 14 A's (10%), 42 B's (30%), 49 C's (35%), 28 D's (20%), and 8 F's (5%).

The null hypothesis (H_0) presumes that the differences occurred by pure chance. If H_0 is demonstrated to be false at an acceptable level of confidence, H_1 is accepted and it must be concluded that an influence other than chance caused the differences in the frequency count.

To determine the odds that the teacher's distribution could occur by chance alone (and was thus not a violation of policy), the department chair applied the chi-square technique. The formula for chi-square is as follows:

$$\chi^2 = \Sigma \left(\frac{(O-E)^2}{E} \right), \tag{13.01}$$

where O = the observed frequency and E = the expected frequency.

To apply this formula, we must find the difference between the observed (teacher's grades) and expected (department policy) frequencies for each category and square each of these differences. Each squared difference is then divided by the expected frequency for its category. Finally we obtain the sum of the squared differences divided by expected frequency. Table 13.1 demonstrates the process.

Table 13.1 is called a one by five (1×5) classification table because the observed frequencies occupy only one row and it has five columns.

In this problem, chi-square = 35.72. The degrees of freedom in chi-square are the number of categories, k (in this case k = number of grade categories), minus one ($df = k - 1$, or $5 - 1 = 4$). Table A.10 in appendix A indicates that for $df = 4$, chi-square must be 13.28 to reach the 99% level of confidence ($p = .01$). Because 35.72 easily exceeds 13.28, the odds that the two distributions differ by chance alone are less than 1 in 100. The department chair rejected H_0 and concluded that the teacher was awarding grades that were significantly higher than permitted by departmental policy.

Table 13.1　Calculation of Chi-Square

Grades	A	B	C	D	F	Total
Observed (teacher's grades)	30	57	32	15	7	141*
Expected (departmental policy)	14	42	49	28	8	141*
$O - E$	16	15	−17	−13	−1	
$(O–E)^2$	256	225	289	169	1	
$(O–E)^2 / E$	18.29	5.36	5.90	6.04	.13	35.72

* The sum of the observed and the expected frequencies must be equal.

Chi-Square (Two or More Classifications)

Chi-square may also be used to solve more complicated problems. Our second example uses two classifications in a contingency table, which is a distribution of frequencies into both rows and columns. One category of frequency occupies the rows and the other category occupies the columns. The following problem involves a 2 × 3 (two rows by three columns) contingency table.

An Example From Motor Behavior

A teacher wanted to know which of two methods of teaching the front crawlstroke was more effective in producing swimmers with good stroke form. The first method was the traditional part method, in which the kick was taught first, then the arm stroke, then the breathing, and finally all the parts were put together as a complete stroke. The second method was the whole method, in which after the prone float and glide had been learned, the students were introduced to all parts of the stroke simultaneously and practiced the entire stroke for the remainder of the semester.

After teaching 55 students the part method and 52 students the whole method for 15 weeks, twice weekly for 1 hour each class, the teacher asked another swimming expert to classify each student's stroke form into one of three categories: good, average, or poor. The results are shown in table 13.2.

The values in the cells represent the number of students in each class who were classified as good, average, or poor. To obtain this data, a cross tabulation must first be conducted on the raw data. Cross tabulation is simply a process in which data are tabulated into appropriate cells and then summed to determine the total number of subjects per cell. In this example, 15 students in the part instruction classes were determined to have good skill, 21 students in the whole instruction class were classified as good, and so on.

We now ask, because there are more whole method students in the good category, and more part method students in the average and poor categories, is the whole method better or could these frequencies have occurred by chance? We use the chi-square technique to obtain the answer.

First, we must assume that the total group of 107 students represents the best estimate that can be obtained for the expected frequencies in each category. Because there is no control group with which each method can be compared, we calculate how many students would be expected to fall into each category had they all been taught alike by either method (i.e., without the influence of whole or part instruction methods). The totals for each group and for each category must be computed as shown in table 13.3.

The percentage of the total number of students expected in each category is calculated by dividing the total number in that category by the grand total.

Good	$36/107 = 33.64\%$
Average	$46/107 = 42.99\%$
Poor	$25/107 = 23.36\%$

With this information, we can calculate how many students would fall into each category of each group if all students were taught alike. This is the expected

Table 13.2   Chi-Square Contingency Table: 2 × 3 Classification

	Skill categories		
Type of instruction	Good	Average	Poor
Part	15	27	13
Whole	21	19	12

Table 13.3   Determination of Percent of Total

	Skill categories			
Type of instruction	Good	Average	Poor	Total
Part	15	27	13	55
Whole	21	19	12	52
Total	36	46	25	107
% of Grand total	33.64	42.99	23.36	—

frequency for the chi-square calculation. It is determined by multiplying the percentage of students expected in each category by the total number in that group.

For the part group:

Good	$.3364 \times 55 = 18.50$
Average	$.4299 \times 55 = 23.64$
Poor	$.2336 \times 55 = 12.85$

For the whole group:

Good	$.3364 \times 52 = 17.49$
Average	$.4299 \times 52 = 22.35$
Poor	$.2336 \times 52 = 12.15$

An alternate, and shorter, method of determining expected frequencies per cell is to apply the following formula:

$$\text{Expected Frequency} = \frac{(\Sigma \text{ row}) (\Sigma \text{ column})}{N}. \qquad (13.02)$$

For example, the expected frequency for the good students in the part group is $(55 \times 36) / 107 = 18.50$.

The expected frequencies are then entered into a table along with the observed frequencies (table 13.4).

The differences between the observed and expected frequencies are now calculated, squared, divided by the expected value, and summed to compute the chi-square statistic. Table 13.5 demonstrates the process; for this example, chi-square (χ^2) is found to be 2.35.

The degrees of freedom for a chi-square contingency table are number of rows minus one times number of columns minus one:

$$df = (R-1)(C-1). \qquad (13.03)$$

In this example, $df = (2 - 1)(3 - 1) = (1)(2) = 2$.

Table 13.4   Observed and Expected Frequencies

	Good		Average		Poor	
	Observed	Expected	Observed	Expected	Observed	Expected
Part	15	18.50	27	23.64	13	12.85
Whole	21	17.49	19	22.35	12	12.15

Table 13.5   Calculation of Chi-Square's Contingency Table

Observed	Expected	$O - E$	$(O - E)^2$	$(O - E)^2 / E$
15	18.50	−3.50	12.25	.662
21	17.49	3.51	12.32	.702
27	23.64	3.36	11.29	.478
19	22.35	−3.35	11.22	.502
13	12.85	.15	.02	.002
12	12.15	−.15	.02	.002
			Chi-square =	2.35

For $df = 2$, table A.10 in appendix A indicates that a chi-square value of 4.61 is needed for the 90% level of confidence. Because the calculated value for χ^2 (2.35) is less than that, we must accept H_0. We conclude that the group frequencies are not significantly different and that students learn equally well with either method. The small differences noted are the result of chance occurrences.

Limitations of Chi-Square

Chi-square does not apply well to small samples, especially those in a 2×2 table. The total number of frequencies (N) should be at least 20, and the value of each cell in the expected frequencies row should not be less than 1. Also, in the 2×2 table, some statisticians suggest subtracting 0.5 from each of the $(O - E)$ values before they are squared to protect against type I errors (Thomas & Nelson, 1991, p. 198).

The chi-square technique can be very helpful for analyzing frequency counts by categories. Any number of groups or categories can be analyzed with this method. Chi-square is relatively easy to apply and is applicable to many problems in kinesiology.

Rank Order Correlation

Spearman's rank order correlation coefficient is used to determine the relationship between two sets of ordinal data. It is the nonparametric equivalent of Pearson's correlation coefficient. Often data in kinesiology result from experts ranking subjects. For example, a teacher may rank students by skill: 1 (highest skill), 2 (next best), and so forth on down to the last rank (lowest skill). In recreational sports, ladder and round-robin tournaments may result in a rank order of individuals or teams.

Even when data have been collected on a parametric variable, the raw data may be converted to rankings by listing the best score as 1, the next best as 2, and so on.

Ranked data are ordinal and do not meet the criteria for parametric evaluation by Pearson's product moment correlation coefficient. To measure the relationship between rank order scores, we must use Spearman's rank order correlation coefficient (rho, or ρ). The formula for Spearman's rho is

$$\rho = 1 - \frac{6\,\Sigma d^2}{N(N^2 - 1)}, \tag{13.04}$$

where d = the difference between the two ranks for each subject and N = the total number of subjects (i.e., the number of pairs of ranks). The number 6 will always be in the numerator.

The degrees of freedom for rho are the same as for Pearson's r: $df = N_{pairs} - 2$. The significance of rho is determined by looking up the value for the appropriate degrees of freedom in table A.11 in appendix A.

An Example From Physical Education

We will use the tennis example introduced at the beginning of the chapter to demonstrate how to apply Spearman's rho. Recall that the students were ranked from highest (1) to lowest (25) on the serving test. These ranks were then compared to the final placements on the ladder tournament. The results are presented in table 13.6.

When two scores are tied in rank data, each is given the mean of the two ranks. The next rank is eliminated to keep N consistent. For example, if 2 subjects tie for 4th place, each is given a rank of 4.5, and the next subject is ranked 6. In table 13.6, there are no ties because a ladder tournament in tennis does not permit ties; one person must win. The instructor also ranked the students on serving skills without permitting ties.

The difference between each student's rank in the ladder tournament and rank in serving ability was determined, squared, and summed. These values are also shown in table 13.6. The signs of the difference scores are not critical because all signs become positive when the differences are squared. With these values we can calculate rho by applying equation 13.04:

$$\rho = 1 - \frac{6\,(362)}{25\,(25^2 - 1)} = 1 - .14 = .86.$$

Table A.11 indicates that for $df = 23$ a value of .54 is needed to reach $p = .01$. The obtained value, .86, is greater than .54, so H_0 is rejected, and it is concluded that tennis serving ability and the ability to win tournament games are related.

Remember, a significant correlation does not prove that being a good server is the cause of winning the game. Many factors are involved in successful tennis ability; serving is just one of them. Other factors related to both serving and winning may cause the relationship.

Table 13.6   Spearman's Rank Order Correlation Coefficient

Subject	Place on ladder	Rank on serve	d	d^2
1	8	10	−2	4
2	16	17	−1	1
3	7	4	3	9
4	24	25	−1	1
5	2	5	−3	9
6	15	9	6	36
7	1	3	−2	4
8	23	24	−1	1
9	6	6	0	0
10	22	20	2	4
11	17	13	4	16
12	5	7	−2	4
13	9	18	−9	81
14	14	19	−5	25
15	25	21	4	16
16	3	1	2	4
17	18	15	3	9
18	13	12	1	1
19	10	16	−6	36
20	19	11	8	64
21	4	2	2	4
22	21	23	−2	4
23	12	8	4	16
24	20	22	−2	4
25	11	14	−3	9
				$\Sigma d^2 =$ 362

Mann-Whitney U Test

The Mann-Whitney U test is used to determine the significance of the difference between rankings of two groups of subjects who have been ranked on the same variable. The U value indicates whether one group ranks significantly higher than the other. It is the nonparametric equivalent of an independent, two-group t test. It may be used instead of the t test when the assumptions of the t test cannot be met, such as when the data are ordinal or highly skewed.

When interval or ratio data are highly skewed, we may want to create one rank order list (on the dependent variable) for all subjects from both groups. For example, the highest scoring person from both groups is ranked 1, the second highest scorer is ranked 2, and so on until all subjects in both groups have been ranked. Then we compare the ranks in group 1 to the ranks in group 2 using the Mann-Whitney U test.

All subjects from both groups are ranked on the same variable and placed in order from highest to lowest, and the subjects in group 1 and group 2 are then listed by their ranks. The sums of the ranks for each group are compared to determine whether the median rankings between the groups differ by more than would be expected by chance alone. The formulas in equations 13.05 and 13.06 are modified from Bruning and Kintz (1977, p. 224).

The formula for U_1 is

$$U_1 = n_1 n_2 + \left(\frac{n_1 (n_1 + 1)}{2} \right) - \Sigma R_1 , \tag{13.05}$$

and the formula for U_2 is

$$U_2 = n_1 n_2 - U_1. \tag{13.06}$$

In equations 13.05 and 13.06, n_1 and n_2 are the number of subjects in each group and ΣR_1 is the sum of the rankings for group 1. (It does not matter which group is designated as group 1.)

When $n_1 + n_2 \geq 20$, a Z score may be computed to determine the significance of the differences in ranks. When $n_1 + n_2 < 20$, we can judge the significance of the smaller U value using tables A.12, A.13, and A.14 in appendix A.

An Example From Motor Learning

A student in a motor learning class was assigned a term project to ascertain if gymnasts had better balance skills than the general population. The student measured 10 gymnasts and 15 nongymnasts on the Bass stick test for upright balance. The test results of all 25 subjects were then ranked (1 for best, 25 for worst) on a single list. The results are shown in table 13.7. For these data, U_1 and U_2 are computed as follows:

$$U_1 = (10)(15) + \frac{10(10+1)}{2} - 103 = 102$$

$$U_2 = (10)(15) - 102 = 48.$$

Table 13.7   Balance Rankings

Gymnasts	Nongymnasts
1	3
2	5
4	6
7	10
8	12
9	13
11	14
17	15
21	16
23	18
	19
	20
	22
	24
	25
$\Sigma R_1 = 103$	$\Sigma R_2 = 222$

Because $n_1 + n_2 \geq 20$, a Z score is used to determine significance. The formula to compute Z for either group is

$$Z = \frac{U - \dfrac{n_1 n_2}{2}}{\sqrt{\dfrac{n_1 n_2 \, (n_1 + n_2 + 1)}{12}}}. \qquad (13.07)$$

The number 12 will always be in the denominator within the square root sign.

It does not matter which U value is used to calculate Z. The Z values for U_1 and U_2 will have the same absolute value: One will be positive, and one will be negative. We shall use U_1:

$$Z_1 = \frac{102 - \dfrac{(10)(15)}{2}}{\sqrt{\dfrac{(10)(15)(10 + 15 + 1)}{12}}} = 1.50.$$

Because we are testing H_0, we use a two-tailed test to interpret Z. For a two-tailed test, we need a Z value $= 1.65$ for $p = .10$, $Z = 1.96$ for $p = .05$, and $Z = 2.58$ for $p = .01$ (see table A.1 in appendix A). In this problem, Z does not reach the limits for $p = .10$, so we accept H_0 and conclude that there is no significant difference between gymnasts and nongymnasts in balance ability as measured by the Bass stick test.

Comparing Groups With Small Values of N

If $n_1 + n_2 < 20$, the Z test may be biased, so a table of U (tables A.12–A.14 in appendix A) must be used to determine the significance of U. The critical U value is found in the table and compared to the smaller calculated U. If the smaller U value is equal to or less than the table value, the rank difference is significant. In the balance problem, the smaller U (U_2) is 48. Table A.12 shows that for $p = .10$ (a two-tailed test with $n_1 = 10$, and $n_2 = 15$), the smaller U must be 44 or less. Because $U_2 = 48$, the difference between the ranks is not significant. This agrees with our conclusion based on Z.

Kruskal-Wallis ANOVA for Ranked Data

If data are ranked and there are more than two groups, a nonparametric procedure analogous to simple ANOVA is available called the Kruskal-Wallis ANOVA for ranked data. This procedure produces an H value that, when $N > 5$, approximates the chi-square distribution. Once we have calculated H, we can determine its significance by using the chi-square table A.10 in appendix A for $df = k - 1$, where $k =$ the number of groups to be ranked.

An Example From Athletic Training

Athletic trainers and coaches want to return athletes to competitive condition as soon as possible after a debilitating injury. Anterior cruciate ligament (ACL) tears repaired with surgery require extensive rehabilitation. An athletic trainer wanted to know if accelerated rehabilitation (closed kinetic chain activities using weight-bearing exercises) was superior to normal rehabilitation activities (knee extension and flexion exercises) as compared to no special rehabilitation exercises (control).

Eighteen subjects, each of whom had undergone recent ACL reconstruction with the patellar tendon replacement technique, were selected and divided into three groups: control, normal, and accelerated. After 6 months, three orthopedic physicians evaluated each subject and jointly ranked all 18 according to their level of rehabilitation. Following are the rankings classified according to the type of rehabilitation technique (see table 13.8). Ties are given the average of the two tied ranks.

Table 13.8   Rankings According to Rehabilitation Technique

	Control (R_1)	Normal (R_2)	Accelerated (R_3)
	15	13	3.5
	11	8	16.5
	18	7	1
	16.5	3.5	5
	12	14	6
	10	9	2
Sums	82.5	54.5	34.0
Means	13.75	9.08	5.67

Clearly, there are differences in the sums and the means. The question we must ask is, are the differences large enough to be attributed to the treatment effects, or are they chance differences that we would expect to occur even if the treatments had no effect?

To solve this problem we apply the following formula for the Kruskal-Wallis H value where N = total of all subjects in all groups, n = subjects per group, and k = number of groups.

$$H = \left[\frac{12}{N(N+1)} \right] \left[\frac{\Sigma R_1^2}{n_1} + \frac{\Sigma R_2^2}{n_2} + \ldots \frac{\Sigma R_k^2}{n_k} \right] - 3(N+1) \qquad (13.08)$$

Substituting values from the example in table 13.8, we find,

$$H = \left[\frac{12}{18(18+1)} \right] \left[\frac{82.5^2}{6} + \frac{54.5^2}{6} + \frac{34.0^2}{6} \right] - 3(18+1)$$

$$H = [.035][1822.09] - 57$$

$$H = 6.77$$

From the chi-square table A.10, the critical value for $p = .05$ for $df = 3 - 1 = 2$ is 5.99. Since our obtained value of 6.77 exceeds 5.99, we reject the null hypothesis and conclude that there are significant differences somewhere among the three groups.

The formula for H assumes that no ties have occurred. If more than a few ties occur, a correction for H has been suggested by Spence and others (1968, p. 217). There is usually little practical value in calculating H_c unless the number of ties is large and the value of H is close to the critical value. The correction formula rarely changes the conclusion.

$$H_C = \frac{H}{1 - \frac{(t_1^3 - t_1 + t_2^3 - t_2 + \ldots t_k^3 - t_k)}{N^3 - N}} \tag{13.09}$$

where t = the number of scores tied at a given rank, and k = the number of times ties occur. In table 13.8, there are 2 scores tied at 3.5, and 2 scores tied at 16.5, therefore, $t_1 = 2$, $t_2 = 2$, and $k = 2$.

In our example:

$$H_C = \frac{6.77}{1 - \frac{(2^3 - 2) + (2^3 - 2)}{18^3 - 18}} = \frac{6.77}{.9979} = 6.78$$

To differentiate among the groups, we can calculate the standard error of the difference for any two values using a procedure suggested by Thomas and Nelson (1996, p. 205).

$$SE = \sqrt{[N(N+1)/12][2/n]} \tag{13.10}$$

When ns are unequal, use $1/n_1 + 1/n_2$ in place of $2/n$, and calculate a separate SE for each pair of groups to be compared.

In our example,

$$SE = \sqrt{[18(18+1)/12][2/6]}$$

$$SE = 3.08$$

Because we are making three comparisons, we must use a Bonferroni adjustment (i.e., divide by 3) of our rejection alpha value of .05 (.05/3 = .017). Now we look for pairwise differences at $p = .017$. This protects against type I errors.

For a two-tailed test at $p = .05$, we expect 2.5% of the area under the normal curve to be in each tail. Using the Bonferroni correction to the p value for three comparisons results in 2.5/3 = .83% at each end of the curve. This leaves 50 – .83, or 49.17% of the curve between the mean and the critical value. Using table A.1 we note that the Z score for 49.17% under the curve is ± 2.39. Multiplying our standard error value of 3.08 × ± 2.39 = ± 7.36 gives us the critical value for pairwise comparisons at $p = .017$. To apply this value, it is helpful to create a mean difference table (see table 13.9).

Therefore, we conclude that the accelerated group is significantly different from the control group at $p < .017$, but the normal group is not different from either the control or the accelerated group.

Table 13.9   Mean Differences Among Groups

Group	Control	Normal	Accelerated
Control	0.00	4.67	8.08*
Normal		0.00	3.41
Accelerated			0.00

* Any value in the table that exceeds 7.36 is significant at $p < .017$.

Friedman's Two-Way ANOVA by Ranks

When subjects are measured three or more times using ranked data, or if interval or ordinal data are converted to ranks, a nonparametric procedure similar to repeated measures ANOVA is used. Friedman's two-way ANOVA by ranks computes a chi-square value for the differences between the sum of the ranks for each repeated measure.

An Example From Physical Education

A researcher wanted to know if physical education was judged by students to be more or less popular than selected academic classes. Using elementary school students as subjects, the researcher asked 10 students to rank physical education, math, and English according to how well they liked the classes with 1 representing the most-liked class, 2 the one in the middle, and 3 the least-liked class. Table 13.10 presents the fabricated results of the hypothetical survey.

Friedman's formula to compute chi-square among the sums of the ranks is presented below:

$$X_R^2 = \left[\frac{12}{Nk(k+1)} \right] \left[\Sigma R_1^2 + \Sigma R_2^2 + \dots \Sigma R_k^2 \right] - 3N(k+1) \qquad (13.11)$$

where N = number of subjects, and k = number of repeated measures.

Substituting values from table 13.10, we compute:

$$X_R^2 = \left[\frac{12}{10(3)(3+1)} \right] \left[17^2 + 25^2 + 20^2 \right] - 3(10)(3+1)$$

$$X_R^2 = [.1][1314] - 120 = 11.4$$

Table 13.10 Results of Hypothetical Study

Student	Physical education	Math	English
A	1	3	2
B	2	3	1
C	3	2	1
D	1	2	3
E	1	3	2
F	2	3	2
G	1	2	3
H	2	3	2
I	1	3	2
J	3	1	2
Totals	17	25	20

Using $df = k - 1 = 3 - 1 = 2$, and $p = .01$, we note from table A.10 that the critical chi-square = 9.21. Since the obtained value (11.4) is greater than the critical value, we reject the null hypothesis and conclude at $p < .01$ that students like physical education best.

This study could be conducted with two groups of students, one group taught by the classroom teacher and another group taught by a physical education specialist to determine if there are significant differences between the rankings of the classes under these two conditions.

Summary

When data that do not meet the assumptions of parametric statistics are collected, alternate techniques must be used to reach conclusions about the relationships or the differences among the variables.

Chi-square, which can be used in both single and double classification problems, is the appropriate technique to use to analyze differences in frequency counts of nominal data. Spearman's rank order correlation coefficient may be used to determine the relationship between two variables of ranked data. It is interpreted in a manner similar to Pearson's correlation coefficient for parametric data.

The Mann-Whitney U test can be used to determine the significance of the difference between ordinal rankings of two groups of subjects on the same variable. It is the nonparametric equivalent of an independent t test. Kruskal-Wallis

ANOVA by ranks is used to compare the ranking of three or more independent groups. It is similar to simple ANOVA. Friedman's two way ANOVA by ranks is similar to repeated measures ANOVA. It determines the significance of the difference between ranks on the same subjects.

Problems to Solve

1. A principal in a high school asked a random sample of students if they thought an equal amount of money should be allocated to the boys and girls athletic programs. The boys responded 27 yes and 42 no. The girls responded 35 yes and 36 no. If the total group is the best estimate of the expected response, is there a significant difference between the opinions of the boys and the girls?

2. The following listing shows the order of finish of 10 gymnasts in the compulsory and optional competitions for the all-around event. What is the correlation between the two rankings?

Gymnast	Compulsory	Optional
1	8	6
2	5	7
3	4	5
4	6	4
5	9	9
6	2	1
7	7	8
8	10	10
9	3	2
10	1	3

3. In the national collegiate football rankings, the Eastern teams seemed to have an advantage. The following listing shows the rankings for the top 25 teams by area of the country. Is there a significant difference in favor of the East?

    West—5, 6, 8, 12, 15, 19, 20, 24, 25
    East—1, 2, 3, 4, 7, 9, 10, 11, 13, 14, 16, 17, 18, 21, 22, 23

4. To determine the effectiveness of learning a cartwheel by watching a video, a gymnastics teacher randomly divided 15 novice gymnasts into three groups of 5 students each. One group (control) was shown the cartwheel one time and then tested. A second group (video) watched the video for 5 minutes, then was tested. The third group (live) watched live demonstrations on the cartwheel for 5 minutes then was tested. All 15 subjects were ranked on their ability to perform the cartwheel with the following results (1 is highest rank, 15 is lowest rank).

Control	Video	Live
8	2	1
12	5	3
13	7	4
14	9	6
15	11	10

Is there a significant difference among the groups in their ability to perform the cartwheel?

5. A physical education teacher wanted to know which sport is most popular with middle school students. He asked 14 students to rank football, baseball, and basketball with 1 being liked most and 3 being liked least. Following are the results:

Student	Football	Baseball	Basketball
A	1	2	3
B	2	1	3
C	2	1	3
D	1	3	2
E	2	1	3
F	2	1	3
G	3	2	1
H	2	1	3
I	1	2	3
J	3	1	2
K	3	2	1
L	1	2	3
M	3	1	2
N	2	1	3

Is there a significant difference in the rankings of the three sports?

See appendix C for answers to the problems.

Key Words in This Chapter

Chi-square

Rank order correlation

Mann-Whitney U test

Kruskal-Wallis ANOVA

Friedman's two-way ANOVA

Appendix A
Statistical Tables

Table A.1 Area Under the Normal Curve

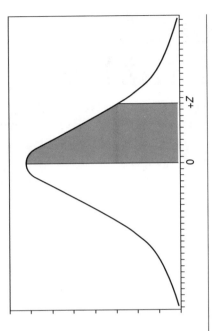

Z	.00	.01	.02	.03	.04	.05	.06	.07	.08	.09	Z
0.0	00.00	00.40	00.80	01.20	01.60	01.99	02.39	02.79	03.19	03.59	0.0
0.1	03.98	04.38	04.78	05.17	05.57	05.96	06.36	06.75	07.14	07.53	0.1
0.2	07.93	08.32	08.71	09.10	09.48	09.87	10.26	10.64	11.03	11.41	0.2
0.3	11.79	12.17	12.55	12.93	13.31	13.68	14.06	14.43	14.80	15.17	0.3
0.4	15.54	15.91	16.28	16.64	17.00	17.36	17.72	18.08	18.44	18.79	0.4
0.5	19.15	19.50	19.85	20.19	20.54	20.88	21.23	21.57	21.90	22.24	0.5
0.6	22.57	22.91	23.24	23.57	23.89	24.22	24.54	24.86	25.17	25.49	0.6
0.7	25.80	26.11	26.42	26.73	27.04	27.34	27.64	27.94	28.23	28.52	0.7
0.8	28.81	29.10	29.39	29.67	29.95	30.23	30.51	30.78	31.06	31.33	0.8
0.9	31.59	31.86	32.12	32.38	32.64	32.89	33.15	33.40	33.65	33.89	0.9
1.0	34.13	34.38	34.61	34.85	35.08	35.31	35.54	35.77	35.99	36.21	1.0
1.1	36.43	36.65	36.86	37.08	37.29	37.49	37.70	37.90	38.10	38.30	1.1
1.2	38.49	38.69	38.88	39.07	39.25	39.44	39.62	39.80	39.97	40.15	1.2

Z	.00	.01	.02	.03	.04	.05	.06	.07	.08	.09	Z
1.3	40.32	40.49	40.66	40.82	40.99	41.15	41.31	41.47	41.62	41.77	1.3
1.4	41.92	42.07	42.22	42.36	42.51	42.65	42.79	42.92	43.06	43.19	1.4
1.5	43.32	43.45	43.57	43.70	43.82	43.94	44.06	44.18	44.29	44.41	1.5
1.6	44.52	44.63	44.74	44.84	44.95	45.05	45.15	45.25	45.35	45.45	1.6
1.7	45.54	45.64	45.73	45.82	45.91	45.99	46.08	46.16	46.25	46.33	1.7
1.8	46.41	46.49	46.56	46.64	46.71	46.78	46.86	46.93	46.99	47.06	1.8
1.9	47.13	47.19	47.26	47.32	47.38	47.44	47.50	47.56	47.61	47.67	1.9
2.0	47.72	47.78	47.83	47.88	47.93	47.98	48.03	48.08	48.12	48.17	2.0
2.1	48.21	48.26	48.30	48.34	48.38	48.42	48.46	48.50	48.54	48.57	2.1
2.2	48.61	48.64	48.68	48.71	48.75	48.78	48.81	48.84	48.87	48.90	2.2
2.3	48.93	48.93	48.98	49.01	49.04	49.06	49.09	49.11	49.13	49.16	2.3
2.4	49.18	49.20	49.22	49.25	49.27	49.29	49.31	49.32	49.34	49.36	2.4
2.5	49.38	49.40	49.41	49.43	49.45	49.46	49.48	49.49	49.51	49.52	2.5
2.6	49.53	49.55	49.56	49.57	49.59	49.60	49.61	49.62	49.63	49.64	2.6
2.7	49.65	49.66	49.67	49.68	49.69	49.70	49.71	49.72	49.73	49.74	2.7
2.8	49.74	49.75	49.76	49.77	49.77	49.78	49.79	49.79	49.80	49.81	2.8
2.9	49.81	49.82	49.82	49.83	49.84	49.84	49.85	49.85	49.86	49.86	2.9
3.0	49.87	49.87	49.87	49.88	49.88	49.89	49.89	49.89	49.90	49.90	3.0
3.1	49.90	49.91	49.91	49.91	49.92	49.92	49.92	49.92	49.93	49.93	3.1
3.2	49.93	49.93	49.94	49.94	49.94	49.94	49.94	49.95	49.95	49.95	3.2
3.3	49.95	49.95	49.95	49.96	49.96	49.96	49.96	49.96	49.96	49.97	3.3
3.4	49.97	49.97	49.97	49.97	49.97	49.97	49.97	49.97	49.97	49.98	3.4
3.5	49.99767										
4.0	49.99968										
4.5	49.99997										
5.0	50.00000										

From *Statistical Tables* by F. James Rohlf and Robert R. Sokal. Copyright © 1969 by W.H. Freeman and Company. Reprinted with permission.

Table A.2 Values of the Correlation Coefficient (r)

df	.10	.05	.01
1	.9877	.9969	.9999
2	.900	.950	.990
3	.805	.878	.959
4	.729	.811	.917
5	.669	.754	.875
6	.621	.707	.834
7	.582	.666	.798
8	.549	.632	.765
9	.521	.602	.735
10	.497	.576	.708
11	.476	.553	.684
12	.457	.532	.661
13	.441	.514	.641
14	.426	.497	.623
15	.412	.482	.606
16	.400	.468	.590
17	.389	.456	.575
18	.378	.444	.561
19	.369	.433	.549
20	.360	.423	.537
25	.323	.381	.487
30	.296	.349	.449
35	.275	.325	.418
40	.257	.304	.393
45	.243	.288	.372
50	.231	.273	.354
60	.211	.250	.325
70	.195	.232	.302
80	.183	.217	.283
90	.173	.205	.267
∞	.164	.195	.254

Table A.3 Values for Student's t Distribution

	Two-tailed test				One-tailed test		
df	.10	.05	.01	df	.10	.05	.01
1	6.314	12.706	63.657	1	3.078	6.314	31.821
2	2.920	4.303	9.925	2	1.886	2.920	6.965
3	2.353	3.182	5.841	3	1.638	2.353	4.541
4	2.132	2.776	4.604	4	1.533	2.132	3.747
5	2.015	2.571	4.032	5	1.476	2.015	3.365
6	1.943	2.447	3.707	6	1.440	1.943	3.143
7	1.895	2.365	3.499	7	1.415	1.895	2.998
8	1.860	2.306	3.355	8	1.397	1.860	2.896
9	1.833	2.262	3.250	9	1.383	1.833	2.821
10	1.812	2.228	3.169	10	1.372	1.812	2.764
11	1.796	2.201	3.106	11	1.363	1.796	2.718
12	1.782	2.179	3.055	12	1.356	1.782	2.681
13	1.771	2.160	3.012	13	1.350	1.771	2.650
14	1.761	2.145	2.977	14	1.345	1.761	2.624
15	1.753	2.131	2.947	15	1.341	1.753	2.602
16	1.746	2.120	2.921	16	1.337	1.746	2.583
17	1.740	2.110	2.898	17	1.333	1.740	2.567
18	1.734	2.101	2.878	18	1.330	1.734	2.552
19	1.729	2.093	2.861	19	1.328	1.729	2.539
20	1.725	2.086	2.845	20	1.325	1.725	2.528
21	1.721	2.080	2.831	21	1.323	1.721	2.518
22	1.717	2.074	2.819	22	1.321	1.717	2.508
23	1.714	2.069	2.807	23	1.319	1.714	2.500
24	1.711	2.064	2.797	24	1.318	1.711	2.492
25	1.708	2.060	2.787	25	1.316	1.708	2.485
26	1.706	2.056	2.779	26	1.315	1.706	2.479

(continued)

Table A.3 *(continued)*

	Two-tailed test				One-tailed test		
df	.10	.05	.01	*df*	.10	.05	.01
27	1.703	2.052	2.771	27	1.314	1.703	2.473
28	1.701	2.048	2.763	28	1.313	1.701	2.467
29	1.699	2.045	2.756	29	1.311	1.699	2.462
30	1.697	2.042	2.750	30	1.310	1.697	2.457
40	1.684	2.021	2.704	40	1.303	1.684	2.423
60	1.671	2.000	2.660	60	1.296	1.671	2.390
120	1.658	1.980	2.617	120	1.289	1.658	2.358
∞	1.645	1.960	2.576	∞	1.282	1.645	2.326

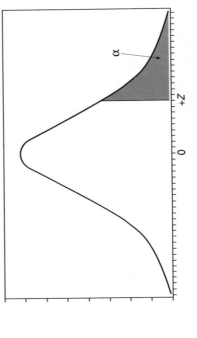

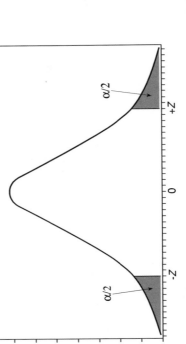

From *Biometrika Tables for Statisticians* (Vol. I) (3rd ed.) by E.S. Pearson and H.O. Hartley (Eds.), 1966, London: Biometrika Trustees. Copyright 1966 by Biometrika Trustees. Reprinted by permission of the Biometrika Trustees.

248

Table A.4 Values of the F Distribution ($p = .10$)

df_E	df_B									
	1	2	3	4	5	6	7	8	9	10
1	39.86	49.50	53.59	55.83	57.24	58.20	58.91	59.44	59.86	60.19
2	8.53	9.00	9.16	9.24	9.29	9.33	9.35	9.37	9.38	9.39
3	5.54	5.46	5.39	5.34	5.31	5.28	5.27	5.25	5.24	5.23
4	4.54	4.32	4.19	4.11	4.05	4.01	3.98	3.95	3.94	3.92
5	4.06	3.78	3.62	3.52	3.45	3.40	3.37	3.34	3.32	3.30
6	3.78	3.46	3.29	3.18	3.11	3.05	3.01	2.98	2.96	2.94
7	3.59	3.26	3.07	2.96	2.88	2.83	2.78	2.75	2.72	2.70
8	3.46	3.11	2.92	2.81	2.73	2.67	2.62	2.59	2.56	2.54
9	3.36	3.01	2.81	2.69	2.61	2.55	2.51	2.47	2.44	2.42
10	3.29	2.92	2.73	2.61	2.52	2.46	2.41	2.38	2.35	2.32
11	3.23	2.86	2.66	2.54	2.45	2.39	2.34	2.30	2.27	2.25
12	3.18	2.81	2.61	2.48	2.39	2.33	2.28	2.24	2.21	2.19
13	3.14	2.76	2.56	2.43	2.35	2.28	2.23	2.20	2.16	2.14
14	3.10	2.73	2.52	2.39	2.31	2.24	2.19	2.15	2.12	2.10
15	3.07	2.70	2.49	2.36	2.27	2.21	2.16	2.12	2.09	2.06
16	3.05	2.67	2.46	2.33	2.24	2.18	2.13	2.09	2.06	2.03
17	3.03	2.64	2.44	2.31	2.22	2.15	2.10	2.06	2.03	2.00
18	3.01	2.62	2.42	2.29	2.20	2.13	2.08	2.04	2.00	1.98
19	2.99	2.61	2.40	2.27	2.18	2.11	2.06	2.02	1.98	1.96
20	2.97	2.59	2.38	2.25	2.16	2.09	2.04	2.00	1.96	1.94
21	2.96	2.57	2.36	2.23	2.14	2.08	2.02	1.98	1.95	1.92
22	2.95	2.56	2.35	2.22	2.13	2.06	2.01	1.97	1.93	1.90
23	2.94	2.55	2.34	2.21	2.11	2.05	1.99	1.95	1.92	1.89
24	2.93	2.54	2.33	2.19	2.10	2.04	1.98	1.94	1.91	1.88
25	2.92	2.53	2.32	2.18	2.09	2.02	1.97	1.93	1.89	1.87
26	2.91	2.52	2.31	2.17	2.08	2.01	1.96	1.92	1.88	1.86
27	2.90	2.51	2.30	2.17	2.07	2.00	1.95	1.91	1.87	1.85
28	2.89	2.50	2.29	2.16	2.06	2.00	1.94	1.90	1.87	1.84
29	2.89	2.50	2.28	2.15	2.06	1.99	1.93	1.89	1.86	1.83
30	2.88	2.49	2.28	2.14	2.05	1.98	1.93	1.88	1.85	1.82
40	2.84	2.44	2.23	2.09	2.00	1.93	1.87	1.83	1.79	1.76
60	2.79	2.39	2.18	2.04	1.95	1.87	1.82	1.77	1.74	1.71
120	2.75	2.35	2.13	1.99	1.90	1.82	1.77	1.72	1.68	1.65
∞	2.71	2.30	2.08	1.94	1.85	1.77	1.72	1.67	1.63	1.60

From *Biometrika Tables for Statisticians* (Vol. I) (3rd ed.) by E.S. Pearson and H.O. Hartley (Eds.), 1966, London: Biometrika Trustees. Copyright 1966 by Biometrika Trustees. Reprinted by permission of the Biometrika Trustees.

Table A.5   Values of the *F* Distribution (*p* = .05)

df_E	df_B									
	1	2	3	4	5	6	7	8	9	10
1	161.4	199.5	215.7	224.6	230.2	234.0	236.8	238.9	240.5	241.9
2	18.51	19.00	19.16	19.25	19.30	19.33	19.35	19.37	19.38	19.40
3	10.13	9.55	9.28	9.12	9.01	8.94	8.89	8.85	8.81	8.79
4	7.71	6.94	6.59	6.39	6.26	6.16	6.09	6.04	6.00	5.96
5	6.61	5.79	5.41	5.19	5.05	4.95	4.88	4.82	4.77	4.74
6	5.99	5.14	4.76	4.53	4.39	4.28	4.21	4.15	4.10	4.06
7	5.59	4.74	4.35	4.12	3.97	3.87	3.79	3.73	3.68	3.64
8	5.32	4.46	4.07	3.84	3.69	3.58	3.50	3.44	3.39	3.35
9	5.12	4.26	3.86	3.63	3.48	3.37	3.29	3.23	3.18	3.14
10	4.96	4.10	3.71	3.48	3.33	3.22	3.14	3.07	3.02	2.98
11	4.84	3.98	3.59	3.36	3.20	3.09	3.01	2.95	2.90	2.85
12	4.75	3.89	3.49	3.26	3.11	3.00	2.91	2.85	2.80	2.75
13	4.67	3.81	3.41	3.18	3.03	2.92	2.83	2.77	2.71	2.67
14	4.60	3.74	3.34	3.11	2.96	2.85	2.76	2.70	2.65	2.60
15	4.54	3.68	3.29	3.06	2.90	2.79	2.71	2.64	2.59	2.54
16	4.49	3.63	3.24	3.01	2.85	2.74	2.66	2.59	2.54	2.49
17	4.45	3.59	3.20	2.96	2.81	2.70	2.61	2.55	2.49	2.45
18	4.41	3.55	3.16	2.93	2.77	2.66	2.58	2.51	2.46	2.41
19	4.38	3.52	3.13	2.90	2.74	2.63	2.54	2.48	2.42	2.38
20	4.35	3.49	3.10	2.87	2.71	2.60	2.51	2.45	2.39	2.35
21	4.32	3.47	3.07	2.84	2.68	2.57	2.49	2.42	2.37	2.32
22	4.30	3.44	3.05	2.82	2.66	2.55	2.46	2.40	2.34	2.30
23	4.28	3.42	3.03	2.80	2.64	2.53	2.44	2.37	2.32	2.27
24	4.26	3.40	3.01	2.78	2.62	2.51	2.42	2.36	2.30	2.25
25	4.24	3.39	2.99	2.76	2.60	2.49	2.40	2.34	2.28	2.24
26	4.23	3.37	2.98	2.74	2.59	2.47	2.39	2.32	2.27	2.22
27	4.21	3.35	2.96	2.73	2.57	2.46	2.37	2.31	2.25	2.20
28	4.20	3.34	2.95	2.71	2.56	2.45	2.36	2.29	2.24	2.19
29	4.18	3.33	2.93	2.70	2.55	2.43	2.35	2.28	2.22	2.18
30	4.17	3.32	2.92	2.69	2.53	2.42	2.33	2.27	2.21	2.16
40	4.08	3.23	2.84	2.61	2.45	2.34	2.25	2.18	2.12	2.08
60	4.00	3.15	2.76	2.53	2.37	2.25	2.17	2.10	2.04	1.99
120	3.92	3.07	2.68	2.45	2.29	2.17	2.09	2.02	1.96	1.91
∞	3.84	3.00	2.60	2.37	2.21	2.10	2.01	1.94	1.88	1.83

From *Biometrika Tables for Statisticians* (Vol. I) (3rd ed.) by E.S. Pearson and H.O. Hartley (Eds.), 1966, London: Biometrika Trustees. Copyright 1966 by Biometrika Trustees. Reprinted by permission of the Biometrika Trustees.

Table A.6   Values of the F Distribution ($p = .01$)

df_E	df_B 1	2	3	4	5	6	7	8	9	10
1	4052	5000	5403	5625	5764	5859	5628	5982	6022	6056
2	98.50	99.00	99.17	99.25	99.30	99.33	99.36	99.37	99.39	99.40
3	34.12	30.82	29.46	28.71	28.24	27.91	27.67	27.49	27.35	27.23
4	21.20	18.00	16.69	15.98	15.52	15.21	14.98	14.80	14.66	14.55
5	16.26	13.27	12.06	11.39	10.97	10.67	10.46	10.29	10.16	10.05
6	13.75	10.92	9.78	9.15	8.75	8.47	8.26	8.10	7.98	7.87
7	12.25	9.55	8.45	7.85	7.46	7.19	6.99	6.84	6.72	6.62
8	11.26	8.65	7.59	7.01	6.63	6.37	6.18	6.03	5.91	5.81
9	10.56	8.02	6.99	6.42	6.06	5.80	5.61	5.47	5.35	5.26
10	10.04	7.56	6.55	5.99	5.64	5.39	5.20	5.06	4.94	4.85
11	9.65	7.21	6.22	5.67	5.32	5.07	4.89	4.74	4.63	4.54
12	9.33	6.93	5.95	5.41	5.06	4.82	4.64	4.50	4.39	4.30
13	9.07	6.70	5.74	5.21	4.86	4.62	4.44	4.30	4.19	4.10
14	8.86	6.51	5.46	5.04	4.69	4.46	4.28	4.14	4.03	3.94
15	8.68	6.36	5.42	4.89	4.56	4.32	4.14	4.00	3.89	3.80
16	8.53	6.23	5.29	4.77	4.44	4.20	4.03	3.89	3.78	3.69
17	8.40	6.11	5.18	4.67	4.34	4.10	3.93	3.79	3.68	3.59
18	8.29	6.01	5.09	4.58	4.25	4.01	3.84	3.71	3.60	3.51
19	8.18	5.93	5.01	4.50	4.17	3.94	3.77	3.63	3.52	3.43
20	8.10	5.85	4.94	4.43	4.10	3.87	3.70	3.56	3.46	3.37
21	8.02	5.78	4.87	4.37	4.04	3.81	3.64	3.51	3.40	3.31
22	7.95	5.72	4.82	4.31	3.99	3.76	3.59	3.45	3.35	3.26
23	7.88	5.66	4.76	4.26	3.94	3.71	3.56	3.41	3.30	3.21
24	7.82	5.61	4.72	4.22	3.90	3.67	3.50	3.36	3.26	3.17
25	7.77	5.57	4.68	4.18	3.85	3.63	3.46	3.32	3.22	3.13
26	7.72	5.53	4.64	4.14	3.82	3.59	3.42	3.29	3.18	3.09
27	7.68	5.49	4.60	4.11	3.78	3.56	3.39	3.26	3.15	3.06
28	7.64	5.45	4.57	4.07	3.75	3.53	3.36	3.23	3.12	3.03
29	7.60	5.42	4.54	4.04	3.73	3.50	3.33	3.20	3.09	3.00
30	7.56	5.39	4.51	4.02	3.70	3.47	3.30	3.17	3.07	2.98
40	7.31	5.18	4.31	3.83	3.51	3.29	3.12	2.99	2.89	2.80
60	7.08	4.98	4.13	3.65	3.34	3.12	2.95	2.82	2.72	2.63
120	6.85	4.79	3.95	3.48	3.17	2.96	2.79	2.66	2.56	2.47
∞	6.63	4.61	3.78	3.32	3.02	2.80	2.64	2.51	2.41	2.32

From *Biometrika Tables for Statisticians* (Vol. I) (3rd ed.) by E.S. Pearson and H.O. Hartley (Eds.), 1966, London: Biometrika Trustees. Copyright 1966 by Biometrika Trustees. Reprinted by permission of the Biometrika Trustees.

Table A.7 Values of the Studentized Range (q) ($p = .10$)

df_E	Number of groups (k)								
	2	3	4	5	6	7	8	9	10
1	8.93	13.4	16.4	18.5	20.2	21.5	22.6	23.6	24.5
2	4.13	5.73	6.77	7.54	8.14	8.63	9.05	9.41	9.72
3	3.33	4.47	5.20	5.74	6.06	6.51	6.81	7.06	7.29
4	3.01	3.98	4.59	5.03	5.39	5.68	5.93	6.14	6.33
5	2.85	3.72	4.26	4.66	4.98	5.24	5.46	5.65	5.82
6	2.75	3.56	4.07	4.44	4.73	4.97	5.17	5.34	5.50
7	2.68	3.45	3.93	4.28	4.55	4.78	4.97	5.14	5.28
8	2.63	3.37	3.83	4.17	4.43	4.65	4.83	4.99	5.13
9	2.59	3.32	3.76	4.08	4.34	4.54	4.72	4.87	5.01
10	2.56	3.27	3.70	4.02	4.26	4.47	4.64	4.78	4.91
11	2.54	3.23	3.66	3.96	4.20	4.40	4.57	4.71	4.84
12	2.52	3.20	3.62	3.92	4.16	4.35	4.51	4.65	4.78
13	2.50	3.18	3.59	3.88	4.12	4.30	4.46	4.60	4.72
14	2.49	3.16	3.56	3.85	4.08	4.27	4.42	4.56	4.68
15	2.48	3.14	3.54	3.83	4.05	4.23	4.39	4.52	4.64
16	2.47	3.12	3.52	3.80	4.03	4.21	4.36	4.49	4.61
17	2.46	3.11	3.50	3.78	4.00	4.18	4.33	4.46	4.58
18	2.45	3.10	3.49	3.77	3.98	4.16	4.31	4.44	4.55
19	2.45	3.09	3.47	3.75	3.97	4.14	4.29	4.42	4.53
20	2.44	3.08	3.46	3.74	3.95	4.12	4.27	4.40	4.51
24	2.42	3.05	3.42	3.69	3.90	4.07	4.21	4.34	4.44
30	2.40	3.02	3.39	3.65	3.85	4.02	4.16	4.28	4.38
40	2.38	2.99	3.35	3.60	3.80	3.96	4.10	4.21	4.32
60	2.36	2.96	3.31	3.56	3.75	3.91	4.04	4.16	4.25
120	2.34	2.93	3.28	3.52	3.71	3.86	3.99	4.10	4.19
∞	2.33	2.90	3.24	3.48	3.66	3.81	3.93	4.04	4.13

From *Biometrika Tables for Statisticians* (Vol. I) (3rd ed.) by E.S. Pearson and H.O. Hartley (Eds.), 1966, London: Biometrika Trustees. Copyright 1966 by Biometrika Trustees. Reprinted by permission of the Biometrika Trustees.

Table A.8   Values of the Studentized Range (q) ($p = .05$)

df_E	Number of groups (k)								
	2	3	4	5	6	7	8	9	10
1	18.0	27.0	32.8	37.1	40.4	43.1	45.4	47.4	49.1
2	6.09	8.3	9.8	10.9	11.7	12.4	13.0	13.5	14.0
3	4.50	5.91	6.82	7.50	8.04	8.48	8.85	9.18	9.46
4	3.93	5.04	5.76	6.29	6.71	7.05	7.35	7.60	7.83
5	3.64	4.60	5.22	5.67	6.03	6.33	6.58	6.80	6.99
6	3.46	4.34	4.90	5.31	5.63	5.89	6.12	6.32	6.49
7	3.34	4.16	4.68	5.06	5.36	5.61	5.82	6.00	6.16
8	3.26	4.04	4.53	4.89	5.17	5.40	5.60	5.77	5.92
9	3.20	3.95	4.42	4.76	5.02	5.24	5.43	5.60	5.74
10	3.15	3.88	4.33	4.65	4.91	5.12	5.30	5.46	5.60
11	3.11	3.82	4.26	4.57	4.82	5.03	5.20	5.35	5.49
12	3.08	3.77	4.20	4.51	4.75	4.95	5.12	5.27	5.40
13	3.06	3.73	4.15	4.45	4.69	4.88	5.05	5.19	5.32
14	3.03	3.70	4.11	4.41	4.64	4.83	4.99	5.13	5.25
15	3.01	3.67	4.08	4.37	4.60	4.78	4.94	5.08	5.20
16	3.00	3.65	4.05	4.33	4.56	4.74	4.90	5.03	5.15
17	2.98	3.63	4.02	4.30	4.52	4.71	4.86	4.99	5.11
18	2.97	3.61	4.00	4.29	4.49	4.67	4.82	4.96	5.07
19	2.96	3.59	3.98	4.25	4.47	4.65	4.79	4.92	5.04
20	2.95	3.58	3.96	4.23	4.45	4.62	4.77	4.90	5.01
24	2.92	3.53	3.90	4.17	4.37	4.54	4.68	4.81	4.92
30	2.89	3.49	3.84	4.10	4.30	4.46	4.60	4.72	4.83
40	2.86	3.44	3.79	4.04	4.23	4.39	4.54	4.63	4.74
60	2.83	3.40	3.74	3.98	4.16	4.31	4.44	4.55	4.65
120	2.80	3.36	3.69	3.92	4.10	4.24	4.36	4.48	4.56
∞	2.77	3.31	3.63	3.86	4.03	4.17	4.29	4.39	4.47

Table A.9   Values of the Studentized Range (q) ($p = .01$)

df_E	Number of groups (k)								
	2	3	4	5	6	7	8	9	10
1	90.0	135	164	186	202	216	227	237	246
2	14.0	19.0	22.3	24.7	26.6	28.2	29.5	30.7	31.7
3	8.26	10.6	12.2	13.3	14.2	15.0	15.6	16.2	16.7
4	6.51	8.12	9.17	9.96	10.6	11.1	11.5	11.9	12.3
5	5.70	6.97	7.80	8.42	8.91	9.32	9.67	9.97	10.2
6	5.24	6.33	7.03	7.56	7.97	8.32	8.61	8.87	9.10
7	4.95	5.92	6.54	7.01	7.37	7.68	7.94	8.17	8.37
8	4.74	5.63	6.20	6.63	6.96	7.24	7.47	7.68	7.87
9	4.60	5.43	5.96	6.35	6.66	6.91	7.13	7.32	7.49
10	4.48	5.27	5.77	6.14	6.43	6.67	6.87	7.05	7.21
11	4.39	5.14	5.62	5.97	6.25	6.48	6.67	6.84	6.99
12	4.32	5.04	5.50	5.84	6.10	6.32	6.51	6.67	6.81
13	4.26	4.96	5.40	5.73	5.98	6.19	6.37	6.53	6.67
14	4.21	4.89	5.32	5.63	5.88	6.08	6.26	6.41	6.54
15	4.17	4.83	5.25	5.56	5.80	5.99	6.16	6.31	6.44
16	4.13	4.78	5.19	5.49	5.72	5.92	6.08	6.22	6.35
17	4.10	4.74	5.14	5.43	5.66	5.85	6.01	6.15	6.27
18	4.07	4.70	5.09	5.38	5.60	5.79	5.94	6.08	6.20
19	4.05	4.67	5.05	5.33	5.55	5.73	5.89	6.02	6.14
20	4.02	4.64	5.02	5.29	5.51	5.69	5.84	5.97	6.09
24	3.96	4.54	4.91	5.17	5.37	5.54	5.69	5.81	5.92
30	3.89	4.45	4.80	5.05	5.24	5.40	5.54	5.65	5.76
40	3.82	4.37	4.70	4.93	5.11	5.27	5.39	5.50	5.60
60	3.76	4.28	4.60	4.82	4.99	5.13	5.25	5.36	5.45
120	3.70	4.20	4.50	4.71	4.87	5.01	5.12	5.21	5.30
∞	3.64	4.12	4.40	4.60	4.76	4.88	4.99	5.08	5.16

From *Biometrika Tables for Statisticians* (Vol. I) (3rd ed.) by E.S. Pearson and H.O. Hartley (Eds.), 1966, London: Biometrika Trustees. Copyright 1966 by Biometrika Trustees. Reprinted by permission of the Biometrika Trustees.

Table A.10 Values of the Chi-Square Distribution (χ^2)

df	.10	.05	.01
1	2.71	3.84	6.63
2	4.61	5.99	9.21
3	6.25	7.81	11.34
4	7.78	9.49	13.28
5	9.24	11.07	15.09
6	10.64	12.59	16.81
7	12.02	14.07	18.48
8	14.36	15.51	20.09
9	14.68	16.92	21.67
10	15.99	18.31	23.21
11	17.28	19.68	24.73
12	18.55	21.03	26.22
13	19.81	22.36	27.69
14	21.06	23.68	29.14
15	22.31	25.00	30.58
16	23.54	26.30	32.00
17	24.77	27.59	33.41
18	25.99	28.87	34.81
19	27.20	30.14	36.19
20	28.41	31.41	37.57
25	34.38	37.65	44.31
30	40.26	43.77	50.89
40	51.81	55.76	63.69
50	63.17	67.50	76.15
60	74.40	79.08	88.80
70	85.53	90.53	100.43
80	96.58	101.88	112.33
90	107.57	113.15	124.12
∞	118.50	124.34	135.81

From *Biometrika Tables for Statisticians* (Vol. I) (3rd ed.) by E.S. Pearson and H.O. Hartley (Eds.), 1966, London: Biometrika Trustees. Copyright 1966 by Biometrika Trustees. Reprinted by permission of the Biometrika Trustees.

Table A.11   Values of the Rank Order Correlation Coefficient (ρ)

df	.10	.05	.01
5	.90		
6	.83	.89	
7	.71	.79	.93
8	.64	.74	.88
9	.60	.68	.83
10	.56	.65	.79
11	.52	.61	.77
12	.50	.59	.75
13	.47	.56	.71
14	.46	.54	.69
15	.44	.52	.66
16	.42	.51	.64
17	.41	.49	.62
18	.40	.48	.61
19	.39	.46	.60
20	.38	.45	.58
21	.37	.44	.56
22	.36	.43	.55
23	.35	.42	.54
24	.34	.41	.53
25	.34	.40	.52
26	.33	.39	.51
27	.32	.38	.50
28	.32	.38	.49
29	.31	.37	.48
∞	.31	.36	.47

From *Non Parametric and Shortcut Statistics* (p. 132) by M.W. Tate and R.C. Clelland, 1957, Danville, IL: Interstate Printers and Publishers. Copyright renewed 1985 by Ruth T. Owen and Richard C. Clelland. Used by permission.

Table A.12 Values of the Mann-Whitney U Distribution Critical Values of U, $p = .05$ (one-tailed), $p = .10$ (two-tailed)

N_1	N_2 9	10	11	12	13	14	15	16	17	18	19	20
1											0	0
2	1	1	1	2	2	2	3	3	3	4	4	4
3	3	4	5	5	6	7	7	8	9	9	10	11
4	6	7	8	9	10	11	12	14	15	16	17	18
5	9	11	12	13	15	16	18	19	20	22	23	25
6	12	14	16	17	19	21	23	25	26	28	30	32
7	15	17	19	21	24	26	28	30	33	35	37	39
8	18	20	23	26	28	31	33	36	39	41	44	47
9	21	24	27	30	33	36	39	42	45	48	51	54
10	24	27	31	34	37	41	44	48	51	55	58	62
11	27	31	34	38	42	46	50	54	57	61	65	69
12	30	34	38	42	47	51	55	60	64	68	72	77
13	33	37	42	47	51	56	61	65	70	75	80	84
14	36	41	46	51	56	61	66	71	77	82	87	92
15	39	44	50	55	61	66	72	77	83	88	94	100
16	42	48	54	60	65	71	77	83	89	95	101	107
17	45	51	57	64	70	77	83	89	96	102	109	115
18	48	55	61	68	75	82	88	95	102	109	116	123
19	51	58	65	72	80	87	94	101	109	116	123	130
20	54	62	69	77	84	92	100	107	115	123	130	138

From the "Critical Values of the U Statistic of the Mann-Whitney Test" by D. Auben, 1953, *Bulletin of the Institute of Educational Research at Indiana University*, **1** (No. 2). Copyright 1953 by Indiana University School of Education. Adapted by permission of Gary Ingersoll.

Table A.13 Values of the Mann-Whitney U Distribution Critical Values of U, $p = .025$ (one-tailed), $p = .05$ (two-tailed)

N_1	N_2											
	9	10	11	12	13	14	15	16	17	18	19	20
1											0	0
2	0	0	1	1	1	1	1	1	2	2	2	2
3	2	3	3	4	4	5	5	6	6	7	7	8
4	4	5	6	7	8	9	10	11	11	12	13	13
5	7	8	9	11	12	13	14	15	17	18	19	20
6	10	11	13	14	16	17	19	21	22	24	25	27
7	12	14	16	18	20	22	24	26	28	30	32	34
8	15	17	19	22	24	26	29	31	34	36	38	41
9	17	20	23	26	28	31	34	37	39	42	45	48
10	20	23	26	29	33	36	39	42	45	48	52	55
11	23	26	30	33	37	40	44	47	51	55	58	62
12	26	29	33	37	41	45	49	53	57	61	65	69
13	28	33	37	41	45	50	54	59	63	67	72	76
14	31	36	40	45	50	55	59	64	67	74	78	83
15	34	39	44	49	54	59	64	70	75	80	85	90
16	37	42	47	53	59	64	70	75	81	86	92	98
17	39	45	51	57	63	67	75	81	87	93	99	105
18	42	48	55	61	67	74	80	86	93	99	106	112
19	45	52	58	65	72	78	85	92	99	106	113	119
20	48	55	62	69	76	83	90	98	105	112	119	127

From the "Critical Values of the U Statistic of the Mann-Whitney Test" by D. Auben, 1953, *Bulletin of the Institute of Educational Research at Indiana University*, **1** (No. 2). Copyright 1953 by Indiana University School of Education. Adapted by permission of Gary Ingersoll.

Table A.14 Values of the Mann-Whitney U Distribution Critical Values of U, $p = .01$ (one-tailed), $p = .02$ (two-tailed)

N_1	N_2 9	10	11	12	13	14	15	16	17	18	19	20
1												
2					0	0	0	0	0	0	1	1
3	1	1	1	2	2	2	3	3	4	4	4	5
4	3	3	4	5	5	6	7	7	8	9	9	10
5	5	6	7	8	9	10	11	12	13	14	15	16
6	7	8	9	11	12	13	15	16	18	19	20	22
7	9	11	12	14	16	17	19	21	23	24	26	28
8	11	13	15	17	20	22	24	26	28	30	32	34
9	14	16	18	21	23	26	28	31	33	36	38	40
10	16	19	22	24	27	30	33	36	38	41	44	47
11	18	22	25	28	31	34	37	41	44	47	50	53
12	21	24	28	31	35	38	42	46	49	53	56	60
13	23	27	31	35	39	43	47	51	55	59	63	67
14	26	30	34	38	43	47	51	56	60	65	69	73
15	28	33	37	42	47	51	56	61	66	70	75	80
16	31	36	41	46	51	56	61	66	71	76	82	87
17	33	38	44	49	55	60	66	71	77	82	88	93
18	36	41	47	53	59	65	70	76	82	88	94	100
19	38	44	50	56	63	69	75	82	88	94	101	107
20	40	47	53	60	67	73	80	87	93	100	107	114

From the "Critical Values of the U Statistic of the Mann-Whitney Test" by D. Auben, 1953, *Bulletin of the Institute of Educational Research at Indiana University,* **1** (No. 2). Copyright 1953 by Indiana University School of Education. Adapted by permission of Gary Ingersoll.

Appendix B
Raw Data

A researcher randomly divided 110 men into two groups. For 6 weeks, the control group ate normally, and the experimental group followed a low-fat diet. At the end of the 6 weeks, the weights of all of the men were recorded as listed below. Four subjects did not complete the diet.

Table B.1   Weight in Pounds

		Control Group		
161	155	164	167	154
168	163	176	158	170
150	171	155	149	186
171	175	169	184	161
164	165	181	168	188
173	172	163	191	183
177	168	173	169	152
154	145	167	157	160
166	159	175	166	179
170	164	147	203	157
162	170	156	164	166

		Experimental Group		
179	160	161	173	181
165	152	150	154	169
204	174	186	161	157
159	156	148	145	150
152	151	144	139	146
164	149	167	188	165
153	198	158	142	154
177	159	147	155	160
155	148	132	140	145
151	150	146	126	136
168				

Table B.2 Softball Throw in Feet

A physical education teacher gave a softball throw for distance test to two classes. Each student threw the ball twice (trial 1 and trial 2). Scores were recorded to the nearest foot.

| | Class 1 Trial | | | Class 2 Trial | |
Subject	1	2	Subject	1	2
1	121	136	1	132	138
	143	149		101	115
	102	96		114	121
	141	162		141	136
5	99	97	5	78	76
	134	99		123	96
	129	135		159	162
	148	141		95	119
	111	106		125	129
10	155	162	10	114	146
	118	105		123	117
	127	110		114	108
	72	79		128	122
	137	161		148	133
15	139	144	15	87	93
	113	119		143	116
	152	126		119	115
	126	126		127	109
	106	84		145	152
20	149	153	20	109	84
	123	92		134	138
	177	175		115	82
	133	126		168	154
	119	87		126	122
25	133	138	25	75	103
	147	116		131	139
	94	105		112	125
	139	135		121	119
	125	143		147	153
30	158	162	30	97	115
	116	146		153	159
	136	132		129	150
	121	118		105	110
	164	159		181	167

Subject	Class 1 Trial			Subject	Class 2 Trial		
	1	2			1	2	
35	103	109		35	123	126	
	176	181			118	100	
	125	108			126	115	
	145	160			134	134	
	119	122			83	126	
40	156	155		40	137	142	
	129	110			119	99	
	186	179			149	135	
	86	84			103	109	
	134	127			122	136	
45	117	138		45	138	129	
	143	145			106	101	
	128	150			128	143	
	109	98			91	102	
	137	134					
50	113	120					

Appendix C
Answers to Problems

1. Measurement is the process of comparing to a standard. This process produces data, which are disorganized. The data are organized, treated, and presented for evaluation through a procedure called statistics. Evaluation determines the worth or value of the data.

2. A variable is a characteristic of a person or object that can assume more than one value. If a characteristic can assume only one value, it is a constant. Independent variables are unrelated to each other (e.g., height and intelligence), whereas dependent variables are related to each other (e.g., body fat and weight).

3. Nominal: Places values into mutually exclusive categories, such as male or female, without qualitative value differences.

   Ordinal: Ordered values such as tallest, next tallest, to shortest.

   Interval: Values on a scale where there is the possibility of negative scores, such as temperature scales or judges' scores.

   Ratio: Values on a scale where there is no possibility of negative scores, such as time (races), distance (height), force (weight), or counting events (heart rate).

4. Statistical inference means to infer characteristics of a population based on a sample taken from that population. A sample is a certain portion or fraction of the population, which is any group of persons, places, or things that have at least one characteristic in common. The sample must be random, meaning everyone or everything in the population must have an equal chance of being selected in the sample. If there are subgroups in the population that are of interest, each subgroup should be sampled so that the total sample has the same representation of subgroups as the population; this process is called stratified random sampling.

5. A parameter is a characteristic of a population, whereas a statistic is a characteristic of a sample taken from a population. Statistics are used to estimate parameters.

6. A theory is a belief about a concept or series of concepts. It is neither right nor wrong when it is conceived, it is just an opinion. However, theories produce hypotheses, which can be tested. When a hypothesis is tested and found to be true, it supports the theory. For example, suppose you believed that distributed practice was better than massed practice for learning to make free throws in basketball. One hypothesis that logically results from the theory is that a group of players who shot 100 free throws in a row during practice would have a lower free throw percentage in the games than a group who shot 100 free throws, 10 at a time with a 5-minute break between each set of 10 shots. If the massed group has a lower percentage in games than the distributed group, the hypothesis is true and the theory is credible.

7. Answers will vary based on individual experience.

8. Some examples are *Research Quarterly for Exercise and Sport, Journal of Applied Biomechanics, Medicine and Science in Sports and Exercise, Journal of Strength and Conditioning Research, Journal of Athletic Training, International Journal of Sports Medicine, Journal of Motor Behavior, Sports Medicine Training and Rehabilitation, Perceptual Motor Skill, Journal of Teaching Physical Education,* and *Journal of Applied Sport Psychology.*

Chapter 2

1. $H = 15.9$, $L = 9.1$, $R = 15.9 - 9.1 = 6.8$.

2. $i = 6.8/15 = .453$; i is rounded to .5.

3.

X	f	Cum. f
15.5-15.9	1	72
15.0-15.4	2	71
14.5-14.9	3	69
14.0-14.4	2	66
13.5-13.9	4	64
13.0-13.4	8	60
12.5-12.9	12	52
12.0-12.4	10	40
11.5-11.9	11	30
11.0-11.4	5	19
10.5-10.9	5	14
10.0-10.4	5	9
9.5-9.9	2	4
9.0-9.4	2	2
	N = 72	

4. Real limits of group 12.0–12.4 are 11.95 – 12.449 . . . .

5. Histogram:

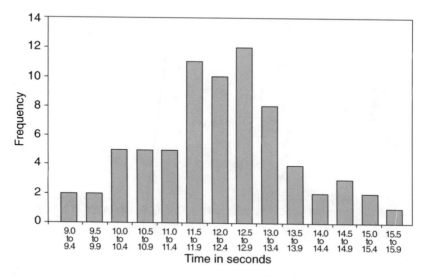

6. Frequency polygon:

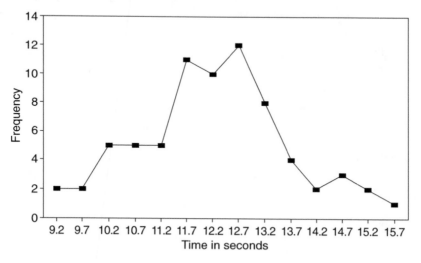

The data appear to be reasonably normal.

7. With more cases, the curve would probably smooth out and approach a more normal shape.

8. Cumulative frequency curve:

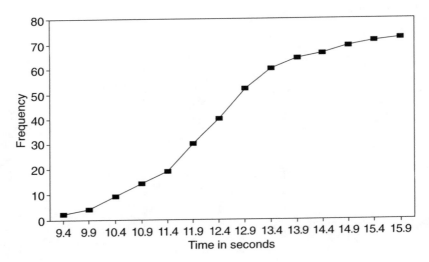

Chapter 3

1. (a) 4/11 × 100 = 36.4%. (b) 8/11 × 100 = 72.7%.
2. (a) .7 × 11 = 7.7 from bottom—nearest score is 10. (b) .45 × 11 = 4.95 from bottom—nearest score is 8.
3. (a) 67/70 × 100 = 95.7%. (b) 4/70 × 100 = 5.7%.
4. (a) .47 × 70 = 32.9 from bottom—score is 10. (b) .80 × 70 = 56 from bottom—score is 13.
5. (a) 13.9%. (b) 30%. (c) 74.6%.
6. (a) 37 sit-ups. (b) 46 sit-ups. (c) 55 sit-ups.

Chapter 4

1. (a) Mean = 9.1. (b) Mean = 12.1.

Chapter 5

1. 4.3.
2. 4.6.
3. $(4.3)^2 = 18.49$.
4. $N = 49$; $\Sigma fX = 1{,}184$; $\Sigma fX^2 = 28{,}796$; $SD = 1.97$.

Chapter 6

1. 13.68.

2. (a) −.79. (b) .62. (c) −.23. All are within 1SD of the mean; (a) and (c) are below, and (b) is above.

3. (a) 21.48%. (b) 73.24%. (c) 40.90%.

4. (a) 42.1. (b) 56.2. (c) 47.7.

5. (a) 3. (b) 6. (c) 5.

6. (a) 3.57. (b) .20. (c) 13.01. As SD increases, SE_M increases. As N increases, SE_M decreases.

7. SE_M = .79; μ = 38.5 ± 1.96 (.79); μ = 38.5 ± 1.55, p = .05. This could be written as 36.95 ≤ μ ≤ 40.05.

Chapter 7

1. (a) +. (b) +. (c) +. (d) −. (e) +. (f) 0. (g) −. (h) −. (i) −.(j) −.

2. (a) r = .833. (b) r = − .833.

3. (a) p < .05. (b) p < .01. (c) not significant.

4. (a) $\bar{X}$ = 24.23 dips; $\bar{Y}$ = 13.47 pull-ups. (b) σ_x = 8.84; σ_y = 4.70. (c) r = .766 for both formulas. (d) Better than 99%, p < .01. (e) 11.75 pull-ups. (f) SE_E = 3.02. (g) 11.75 ± 1.96 (3.02) = 11.75 ± 5.92, or 5.83 to 17.67.

5. Relationship between RT and MT is not significant. r = −.034; RT mean = 240.6; σ_{RT} = 55.204; MT mean = 340.6; σ_{MT} = 89.875.

6. (a) Mile run: Mean = 535.72, σ = 129.68. Max VO_2: Mean = 41.26, σ = 10.02, r = −.94 (b) p < .001. (c) Slope = −.073, Y-intercept = 80.17, SE_E = 3.52. (d) Y_p = −.073 (540) + 80.17 ± 1.96 (3.52), Y_p = 40.75 ± 6.90 ml/kg/min.

Chapter 8

1. (a) 3.81. (b) 1.19.

2. (a) t = −.80, N.S. (b) t = 2.27, df = 23, p < .05.

3. The standard error of the difference is the standard deviation of an infinite set of scores, each of which was derived by computing the difference between the means of two samples randomly drawn from the same population. It is the amount of difference expected between two randomly drawn sample means due to chance alone.

4. No significant difference, $t = .511$. Mean active = 2.282 ($SD = 1.244$), mean passive = 1.966 ($SD = 1.506$), $SE_D = .618$.

5. $SE_D = .066$, $df = 35$, $t = 3.03$, $p < .01$, $\omega^2 = .18$, $ES = .87$. The difference is significant, the effect of being on the cross-country team on stride length is large, and 18% of the variability in stride length can be explained by participation on a cross-country ski team.

6. $SE_{M_1} = 2.84$, $SE_{M_2} = 3.18$, $SE_D = 1.78$, $t = 2.81$. The research hypothesis could be used because it is logical to assume that the dominant hand would be stronger. For $df = 19$, a one-tailed test is significant at $p < .01$, and a two-tailed test is significant at $p < .05$.

7. (a) Males: mean = 20.0, $SD = 3.16$. Females: mean = 21.17, $SD = 2.32$. $SE_D = 1.65$, $t = -.707$, N.S. (b) Accept.

8. (a) Yes. First: mean = 39.8, $SD = 11.86$. Second: mean = 47.8, $SD = 9.83$. $SE_D = 1.50$, $r = .921$, $t = -5.34$. (b) $p < .01$, reject null.

9. Area for $Z_\beta = 40\%$, $Z_\beta = 1.28$. The value of N needed for 90% power at $p = .05$ is approximately 9 subjects per group.

Chapter 9

1. (a) Means: good = 6.00, average = 7.83, poor = 11.00. (b)–(d) ANOVA table:

Source	SS	df	MS	F	p
Between	76.78	2	38.39	15.61	< .01
Within	36.83	15	2.46		
Total	113.61	17			

(e) Mean difference table:

	1	2	3
1	0.00	1.83	5.00*
2		0.00	3.17**
3			0.00

* $p < .01$ for both Scheffé and Tukey.
** $p < .05$ for Scheffé and $p < .01$ for Tukey.

(f) $I_{.05} = 2.46$; $I_{.01} = 3.23$; 1 vs. 3, 2 vs. 3 differ. (g) $HSD_{.05} = 2.35$; $HSD_{.01} = 3.09$; 1 vs. 3, 2 vs. 3 differ. (h) Scheffé and Tukey agree on groups 1 and 3, $p < .01$. Scheffé finds 2 and 3 different at $p < .05$ but Tukey finds 2 and 3 different at $p < .01$. (i) $R^2 = .68$, $\omega^2 = .62$, effect is large.

2. (a) IV = activity level. DV = body fat. (b)

	Inactive	Semi	Normal	Active	Very
Means	27.13	24.97	22.22	16.12	12.35
SD	5.80	4.37	2.73	3.73	1.95

(c) Yes. $F = 14.68$, $p < .001$. (d) Mean difference table:

	1	2	3	4	5
1	0.00	2.16	4.91	11.01*	14.78*
2		0.00	2.75	8.85*	12.62*
3			0.00	6.10	9.87*
4				0.00	3.77
5					0.00

$HSD_{.01} = 8.33$, $*p < .01$

Chapter 10

1. (a) Means = 16.6, 20.8, 25.0, 30.6, 35.4.(b)–(e) ANOVA repeated measures table:

Source	SS	df	MS	F	p
Columns	1127.04	4	281.76	98.52	< .01
Rows	220.64	4	55.16	19.29	< .01
Error	45.76	16	2.86		
Total	1393.44	24			

(f) Yes. $F = 98.52$, $p < .01$, indicates that at least two of the means are significantly different. (g) $HSD_{.01} = 4.15$. Each of the means is significantly different from all of the others. (h) ANOVA repeated measures table for pilot data is given below. $R_1 = .972$, $R_2 = .980$. Pilot data are highly reliable.

Source	SS	df	MS	F	p
Columns	12.96	4	3.24	3.06	.05
Rows	214.16	4	53.54	50.51	.01
Error	17.04	16	1.06		
Total	1393.44	24			

Means = 16.6, 17.2, 17.2, 18.4, 18.4

2. (a)

	Machine 1	Machine 2	Machine 3	Machine 4	Control
Means	623.60	597.70	568.30	586.30	620.50
SD	367.15	392.00	333.73	448.57	395.25

(b) No. $F = .492$, N.S. (c) $p = .744$ by computer analysis. (d) No, when F is not significant, further mean difference comparisons are not justified.

Chapter 13

1. No significant difference. Chi-square = 1.46, N.S. $df = (2 - 1)(2 - 1) = 1$. See tables below for calculations.

	Yes		No		
	O	E	O	E	Total
Boys	27	30.6	42	38.5	69
Girls	35	31.4	36	39.5	71
Total	62	62	78	78	140
% expected	44.3		55.7		

O	E	O–E	$(O–E)^2$	$(O–E)^2 / E$
27	30.6	–3.6	12.96	.42
35	31.4	3.6	12.96	.41
42	38.5	3.5	12.25	.32
36	39.5	–3.5	12.25	.31

Chi-square = 1.46

2. $\rho = .88$, $p < .01$.

3. No, difference is not significant.

$$\Sigma R_{\text{West}} = 134; \; \Sigma R_{\text{East}} = 191.$$

$$U_{\text{West}} = (9)(16) + \left[\frac{9(9+1)}{2} \right] - 134 = 55$$

$$U_{\text{East}} = (9)(16) - 55 = 89$$

$$Z_{\text{West}} = \frac{55 - \dfrac{(9)(16)}{2}}{\sqrt{\dfrac{(9)(16)(9+16+1)}{12}}} = -.96, \text{ N.S.}$$

4. Yes. Live is better than control.

$H = (.05)(1115.2) - 48 = 7.76$, $df = 3 - 1 = 2$, $p < .05$.

Mean Difference Table

	Control	Video	Live
Control	0.0	5.6	7.6*
Video		0.0	2.0
Live			0.0

$SE = 2.83$. Critical difference $= 2.83 \times 2.39 = \pm 6.67$.
* $p = < .017$.

5. Yes. The students like baseball best.

$$X^2 = \left(\frac{12}{168} \right)(2450) - 168 = 7.0$$

$$df = 3 - 1 = 2$$

$$p < .01$$

Glossary

actual mean difference: the difference between the raw score mean values in the numerator of a t test.

alpha (α): (1) the area under the normal curve for rejection of H_0; (2) the probability of chance occurrence.

analysis of covariance (ANCOVA): the adjustment of dependent variable mean values to account for the influence of one or more covariates that are not controlled by the research design.

analysis of variance (ANOVA): an F value that represents the ratio of between-group and within-group variance.

apparent limits: the integer or discrete values listed for each group in a grouped frequency distribution.

best fit line: (1) a line on a scatter plot that best indicates the relationship between the plotted values; (2) a line on a scatter plot that balances the positive and negative residual values so that they sum to zero.

beta weight: a coefficient in multiple regression indicating the weight to be assigned to each independent variable in the prediction equation. Beta weights are in the form of Z scores when the independent variables have been converted to standard scores.

between-group variance: in ANOVA, the deviation of a set of group means from the grand mean.

between-within: a factorial ANOVA comparing independent groups (between) measured two or more times (within), sometimes called a mixed model.

bias: the factors operating on a sample so it is not representative of the population from which it was drawn.

bimodal: a distribution of values with more than one mode.

bivariate regression: regression analysis applied to one independent variable (X) and one dependent variable (Y).

Bonferroni adjustment: an adjustment of the p value (probability of error) to correct for a familywise error rate when making multiple comparisons on the same set of subjects.

ceiling effect: the phenomenon that makes it more difficult to improve as a raw score approaches the maximum possible score.

central limit theorem: a theorem stating that the means of a series of samples randomly selected from a population will be normally distributed even if the population from which they were taken is not normal.

central tendency: values that describe the middle, or central, characteristics of a set of data. The three values of central tendency are the mode, median, and mean.

chi-square: a nonparametric statistical technique for determining the significance of the difference between frequency counts on nominal data.

circularity: the assumption of sphericity applied to the pooled data (across all groups) in a between-within or within-within factorial ANOVA.

coefficient: a known numerical quantity used to explain or modify a variable.

constant: a characteristic of a person, place, or thing that can assume only one value.

continuous variable: a variable that can theoretically assume any level on a continuum of data depending only on the accuracy of the instrument used to measure the variable (i.e., it can be measured to the nearest 10th, 100th, 1,000th, etc., and there are no gaps in the range of the data).

correlation: (1) a numerical coefficient between +1.00 and −1.00 that indicates the extent to which two variables are related or associated; (2) the extent to which the direction and size of deviations from the mean in one variable are related to the direction and size of deviations from the mean in another variable.

criterion variable: another name for the dependent variable.

critical ratio: the ratio or numerical result of a *t* test that must be met to reach a given level of confidence when rejecting the null hypothesis.

cross validation: a confirmation of the accuracy of a bi- or multivariate regression equation by applying the equation to a second independent sample taken from the same population.

cumulative frequency graph: a line graph of ordered scores plotted against the frequency of the scores that fall at or below a given score.

curve: a line on an X-Y plot representing the succession of change on Y as values of X are altered.

curvilinear: a plot of bivariate data where the best-fit line is not straight.

data: information gathered by measurement.

decile: one-tenth of the range of values.

degrees of freedom: the number of values in a data set that are free to vary when restrictions are imposed on the set.

dependent sample: a sample related to another sample and dependent on it. Dependent samples usually contain the same people measured more than once (i.e., pre-post comparisons).

dependent variable: a variable whose value is partially determined by the effects of other variables. It is not free to assume any value. It is usually the variable that is measured in the research design.

descriptive research: an attempt to determine the current state of events or conditions.

descriptive statistics: numerical values that describe a current event or condition.

discrete variable: a variable that is limited in its assessment to certain values, usually integers (i.e., the data is not continuous; there are gaps between values in the range of the data).

discriminate analysis: classification of subjects into like groups based on certain measureable characteristics of the subjects.

effect size: (1) an estimate of the percent of total variance between means that can be attributed to the result of treatment; (2) a measurement of the magnitude of the difference between two mean values independent of sample size.

eigenvalue: a coefficient used in factor analysis that indicates the number of original variables that are associated with a single factor.

eta squared: a measure of effect size in ANOVA; same as R^2.

evaluation: the philosophical process of determining the worth, or value, of the data.

expected mean difference: another term for standard error of the difference.

experimental design: a research process that involves manipulating and controlling certain events or conditions to solve a problem.

external validity: the ability to generalize the results of an experiment to the population from which the samples were drawn.

factor: (1) a component in the design of a study that is combined with other factors to answer multiple questions about the data; (2) a virtual variable that is the result of a combination of two or more variables in a factor analysis design.

factor analysis: a research design that compares two or more variables simultaneously.

factorial ANOVA: analysis of variance performed on more than one factor (e.g., the effects of gender (factor A) and treatment (factor B) analyzed simultaneously). Factorial ANOVA permits evaluation of the interaction of the factors on the dependent variable.

familywise error rate: an inflation of the error rate when making a series or family of comparisons on the same set of subjects, sometimes called inflated alpha.

frequency data: (1) values associated with raw scores indicating the number of subjects who received each raw score; (2) values associated with nominal data indicating the number of subjects classified into each category.

frequency polygon: a line graph of scores plotted against frequency.

Friedman's two-way ANOVA: a nonparametric test similar to repeated measures ANOVA used to determine the significance of the differences among groups of ranked data collected as repeated measures.

graph: (1) a diagrammatic representation of quantities designed to show their relative values; (2) a visual representation of data.

Greenhouse-Geisser adjustment: a conservative adjustment to the degrees of freedom in repeated measures ANOVA to correct for violation of the assumption of sphericity.

grouped frequency distribution: an ordered listing of the values of a variable organized into groups with a frequency column indicating the number of cases included in each group.

hierarchical multiple regression: a multiple regression technique where the investigator specifies the order of inclusion of the independent variables in the solution.

histogram: a graph plotting blocked scores against frequency; commonly known as a bar graph.

historical research: a search of the records of the past in an attempt to determine what happened and why.

homogeneity of covariance: equality of the correlation coefficients among a set of three or more repeated measures on the same subjects.

homogeneity of variance: equality of the variances among a set of two or more measures.

homoscedasticity: a condition where the variance of the residuals of each of the independent variables in multiple regression is equal or nearly so.

Huynh-Feldt adjustment: a liberal adjustment to the degrees of freedom in repeated measures ANOVA to correct for violation of the assumption of sphericity.

hypothesis: a prediction or assumption that can be tested to determine whether or not it is correct.

independent sample: a sample that is unrelated to a second sample. Independent samples usually contain different subjects in each sample.

independent variable: (1) variable that is free to vary and that is not dependent on the influence of another variable; (2) a variable in the research design that is permitted to exert influence over other variables (i.e., the dependent variable) in the study. The independent variable is usually controlled by the research design.

instrument error: bias or error in the data produced by inaccurate or improperly calibrated instruments.

interaction: the combined effect of two or more factors on the dependent variable.

interindividual variability: the amount of variability in measurements that can be attributed to differences between two or more different subjects (people).

internal validity: the ability of the research design to determine that the results are due to the treatment applied.

interquartile range: the range in raw score units from the 25th percentile (Q_1) to the 75th percentile (Q_3).

interval scale: a parametric scale of measurement with equal units or intervals between data points, but which has no absolute zero point (i.e., the Fahrenheit scale of temperature). Ratio comparisons among the data points are not appropriate.

interval size: the numerical size of each group in a group frequency distribution (i.e., the number of data points in the group).

intervening variable: an extraneous variable not controlled by the research design that has an unintended or unknown effect on the dependent variable.

intraclass reliability or intraclass correlation: a method of determining reliability, which is sensitive to both order and magnitude of change in data, based on repeated measures ANOVA. It may be used on three or more repeated measures.

intraindividual variability: the amount of variability in measurements that can be attributed to changes within the individual subjects (people).

investigator error: bias or error in the data produced by the human limitations of the investigator.

J-curve: a curve that results when frequency is high at one end of the scale, decreases rapidly, and reduces to almost zero at the other end of the scale.

kinesiology: the study of the art and science of human movement related to sport, dance, and exercise.

Kruskal-Wallis ANOVA: a nonparametric test similar to ANOVA used to determine the significance of the differences among groups of ranked data.

kurtosis: a measure of the vertical deviation from normality (amount of peakedness or flatness) in the plot of a data set.

leptokurtic: a curve that is more peaked than a mesokurtic curve.

level of confidence: (1) the amount of confidence that can be placed in a conclusion; (2) a value expressed as a percentage that establishes the probability that a statement is correct.

main effects: the F value based on marginal means in a factorial analysis of variance.

Mann-Whitney U test: a nonparametric statistical technique for determining the significance of the difference between rankings of two groups of subjects who have been ranked on the same variable.

marginal means: the average scores across all groups or all trials in a factorial analysis of variance.

mean: the arithmetic average.

mean square: in ANOVA, the average of the squared deviations from a mean of a set of scores.

measurement: the process of comparing to a standard.

median: the 50th percentile, or the score that falls midway in the range of ordered values.

mesokurtic: a typical, bell-shaped, normal curve.

mixed model: a between-within research design in factorial analysis of variance.

mode: the score in a distribution of values that occurs the most often.

multicollinearity: a condition where two or more independent variables in multiple regression are highly correlated with each other.

multiple analysis of variance (MANOVA): the simultaneous analysis of two or more dependent variables in a research design using analysis of variance.

multiple regression: regression analysis applied to more than one independent variable (X_1, X_2, X_3, etc.) and one dependent variable (Y).

multivariate: data sets consisting of measurement of more than one variable per subject.

nominal scale: a nonparametric classification of data based on names of categories and the frequency of occurrence within each category.

nonparametric: data that do not meet the assumption of normality.

normal curve: a curve, which has known characteristics, formed by the bilaterally symmetrical, bell-shaped distribution of values around the mean.

null hypothesis: a hypothesis that predicts the absence of a relationship among subjects or no differences between or among groups of subjects. It is traditionally symbolized as H_0. It is typically the hypothesis that is tested statistically.

objectivity: an appraisal of the amount of bias in the measurement process.

omega squared: a measure of effect size that determines the percent of total variance that may be attributed to treatment.

one-tailed test: a test of the research hypothesis wherein the difference between two mean values is predicted to be significant. It uses only one tail of the normal curve.

ordinal scale: a nonparametric listing of data based on order without consideration of the absolute value of each data point (i.e., a listing from highest to lowest; first, second, third, etc.).

outlier: a value in a data set that lies beyond the limits of the typical scores.

parameter: a characteristic of a population.

parametric: data that meet the assumptions of normality.

partial correlation: the correlation (common variance) between an independent variable and a dependent variable that remains after the influence of a second variable has been removed.

Pearson's product moment correlation coefficient: a correlation coefficient appropriate for use on parametric data.

percentile: a point or position on a continuous scale of 100 theoretical divisions such that a certain fraction of the population of scores lies at or below that point.

platykurtic: a curve that is more flat than a mesokurtic curve.

population: a group of people, places, or things that have at least one common characteristic.

power: the ability of a test to correctly reject a false null hypothesis.

predictor variable: another name for the independent variable.

probability of error: (1) the probability that a statement is incorrect; (2) a value expressed as a decimal that establishes the probability that a statement is incorrect.

quartile: one-fourth of the range of values.

quintile: one-fifth of the range of values.

R squared (R^2): (1) a measure of effect size in ANOVA; (2) the ratio of the between-group variance and the total variance in ANOVA; same as eta squared.

random sample: a sample taken from a population where every member of the population has an equal chance of being selected in the sample.

range: the numerical distance from the highest to the lowest score.

rank order correlation: same as Spearman's rho.

rank order distribution: an ordered listing of data in a single column.

ratio scale: a parametric scale of measurement based on order, with equal units of measurement and an established zero point (i.e., data based on time, distance, force, or counting events).

real limits: the assumed upper and lower values for a group in a grouped frequency distribution that include all possible values on a continuous scale.

rectangular curve: a curve that results when the frequencies of the values in the middle of the data are the same.

regression: (1) a method of predicting values on the Y-variable based on a value on one or more X-variables and the relationship between the variables; (2) a statistical term meaning prediction.

reliability: a measure of the consistency of the data when measurements are taken more than once under the same conditions.

repeated measures: measuring the same set of subjects more than once, as in a pre-post comparison; same as within-subjects design.

research: a special technique for solving problems.

research hypothesis: a hypothesis that typically prompts research. It is usually a prediction that there are significant relationships among subjects or significant differences between or among groups of subjects. It is traditionally symbolized as H_1.

sample: a portion or fraction of a population.

sampling error: the amount of error in the estimate of a population parameter based on a sample.

scatter plot: a plot of bivariate data with one variable on the X-axis and the other on the Y-axis that produces a visual picture of the relationship between the variables.

Scheffé's confidence interval (I): a post hoc test conducted after a significant ANOVA to determine the significance of all possible combinations of cell contrasts.

semi-interquartile range: half the interquartile range.

significant: a statistical term meaning that a relationship or a mean difference is not due solely to chance.

simple effects: the F value based on a single group or a single trial in factorial analysis of variance.

simple frequency distribution: an ordered listing of the values of the variable with a frequency column that indicates the number of cases for each value.

singularity: a condition where two or more independent variables in multiple regression are perfectly correlated ($r = 1.00$) with each other.

skewed: a plot of values that is not normal (i.e., a disproportionate number of subjects fall toward one end of the scale—the curve is not bilaterally symmetrical).

skewness: a measure of lateral deviation from normality (bilateral symmetry) in the plot of a data set.

Spearman's rho: a nonparametric correlation technique for determining the significance of the relationships among ordinal data sets; same as rank order correlation.

sphericity: an assumption that in repeated measures designs the variance among the repeated measures is the same or nearly so (homogeneity of variance), and the relationships or correlations among all of the combinations of the repeated measures are the same or nearly so (homogeneity of covariance).

standard deviation: (1) a measure of the spread, or dispersion, of the values in a parametric data set standardized to the scale of the unit of measurement of the data; (2) the square root of the average of the squared deviations around the mean.

standard error of the difference: (1) the amount of difference between the means of two randomly drawn samples that may be attributed to chance alone; (2) the denominator in the t test.

standard error of the estimate: a numerical value that indicates the amount of error in the prediction of a Y value in bivariate or multivariate regression.

standard error of the mean: the numeric value that indicates the amount of error in the prediction of a population mean.

standard multiple regression: a multiple regression technique where all of the independent variables are entered into the solution in one step.

standard score: a score that is derived from raw data and that has a standard basis for comparison (i.e., it has a known central tendency and a known variability).

stanine: a standard score based on the division of the normal curve into nine sections, each of which is one-half of a standard deviation wide, with a mean of 5 and a range of 1 to 9.

statistic: a characteristic of a sample.

statistical inference: to infer certain characteristics of a population based on random samples taken from that population.

statistics: a mathematical technique by which data are organized, treated, and presented for interpretation and evaluation.

stepdown analysis: a step-by-step process comparing main effects first, simple effects second, and cell contrasts last in a factorial analysis of variance.

stepwise multiple regression: a multiple regression technique where the order of inclusion of the independent variables in the solution is determined according to the amount of explained variance each independent variable can offer.

stratified sample: a series of samples taken from various subgroups of a population so that the subgroups are represented in the total sample in the same proportion that they are found in the population.

sum of squares: the sum of the squares of the deviations from a mean of a set of scores.

T score: a standard score with a mean of 50 and a standard deviation of 10.

terminal statistic: a statistic that does not provide information that can be used in further analysis of the data.

theory: a belief about a concept or series of concepts.

total variance: the sum of between-group and within-group variability.

Tukey's honestly significant difference (*HSD*): a post hoc test conducted after a significant ANOVA to determine the significance of pairwise cell contrasts.

two-tailed test: a test of the null hypothesis wherein a difference between two mean values is predicted to be zero. It uses both tails of the normal curve.

type I error: rejection of the null hypothesis when it is really true.

type II error: acceptance of the null hypothesis when it is really false.

U-shaped curve: a curve that results when the frequencies of the values at the extremes of the scale are higher than the frequency of values in the middle.

validity: the soundness or correctness of a test or instrument in measuring what it is designed to measure (i.e., the truthfulness of the test or instrument).

variability: a measure of the spread or dispersion of a set of data. The four most commonly used measures of variability are range, interquartile range, variance, and standard deviation.

variable: a characteristic of a person, place, or thing that can assume more than one value.

variance: (1) a measure of the spread, or dispersion, of the values in a parametric data set; (2) the average of the squared deviations around the mean.

within-group variance: in ANOVA, the deviation of a set of values from the mean of the group in which they are included.

within-subjects design: measurements within the same set of subjects; same as a repeated measures design.

Y-intercept: the point where the extension of the best fit line intercepts the Y-axis.

Z score: a standard score with a mean of 0 and standard deviation of 1.

References

AndersonBell. (1989). *ABstat users manual*. Parker, CO: Author.

Bruning, J.L., & Kintz, B.L. (1977). *Computational handbook of statistics (2nd ed.)*. Glenview, IL: Scott Foresman.

Cohen, J. (1988). *Statistical power analysis for the behavioral sciences (2nd ed.)*. Hillsdale, NJ: Erlbaum.

Dallal, G.E. (1986). PC Size: Consultant—A program for sample size determinations. *The American Statistician*, 40, 52.

Dixon, W.J. (Ed.) (1990). *BMDP statistical software manual*. Berkeley, CA: University of California Press.

Duoos, B.A. (1984). Fatigue and diagonal stride in cross-country skiing. In *Sport Biomechanics: Proceedings of the International Symposium of Biomechanics in Sports* (p. 219). Del Mar, CA: The Research Center for Sports, and Academic Publishers.

Egstrom, G. (1964). Effects of an emphasis on conceptualizing techniques during early learning of a gross motor skill. *Research Quarterly*, 35, 472-481.

Finney, C. (1985). Further evidence: Employee recreation and improved performance. *Journal of Employee Recreation, Health, and Education*, 28, 8-10.

Franks, B.D., & Huck, S.W. (1986). Why does everyone use the .05 significance level? *Research Quarterly for Exercise and Sport*, 57, 245-249. (Also see *Research Quarterly for Exercise and Sport*, (1987) 58, 81-89, for comments and responses.)

Jones, J.G. (1965). Motor learning without demonstration of physical practice under two conditions of mental practice. *Research Quarterly*, 36, 270.

Kachigan, S. (1986). *Statistical analysis*. New York: Radius Press.

Keppel, G. (1991). *Design and analysis: A researcher's handbook (3rd ed.)*. Englewood Cliffs, NJ: Prentice Hall.

Kotz, S., & Johnson, N.L. (Eds.) (1982). *Encyclopedia of statistical sciences (Vol. 3)*. New York: Wiley.

Leedy, Paul. (1980). *Practical research planning and design (2nd ed.)*. New York: Macmillan.

Oxendine, J.B. (1969). Effect of mental and physical practice on the learning of three motor skills. *Research Quarterly*, 40, 755.

Richardson, A. (1967). Mental practice: A review and discussion. *Research Quarterly*, 38, 95.

Schutz, R.W., & Gessaroli, M.E. (1987). The analysis of repeated measures designs involving multiple dependent variables. *Research Quarterly for Exercise and Sport*, 58, 132-149.

Spence, J.T., Underwood, B.J., Duncan, C.P., & Cotton, J.W. (1968). *Elementary statistics (2nd ed.)*. New York: Appleton-Century-Crofts.

Spiegel, M.R. (1961). *Theory and problems of statistics*. New York: Schaum.

Statistical Analysis Systems Institute. (1985). *SAS users guide: Statistics (Ver. 5 ed.)*. Cary, NC: Author.

Statistical Package for the Social Sciences. (1988). *SPSS-X users guide (3rd ed.)*. Chicago: Author.

Tabachnick, B.G., & Fidell, L.S. (1995). *Using multivariate statistics*. Cambridge, MA: Harper & Row.

Thomas, J.R., & Nelson, J.K. (1996). *Research methods in physical activity (3rd ed.)*. Champaign, IL: Human Kinetics.

Thomas, J.R., Salazar, W., & Landers, D.M. (1991). What is missing in p less than .05 effect size? *Research Quarterly for Exercise and Sport, 62*, 344.

Twinning, W.E. (1949). Mental practice and physical practice in learning a motor skill. *Research Quarterly, 20*, 432.

Vincent, W.J. (1968). Transfer effects between motor skills judged similar in perceptual components. *Research Quarterly, 39*, 380.

Vu, T.Z. (1997). Measurement in Physical Education and Exercise Science, 1, 89.

Wagoner, K.D. (1994). Descriptive discriminant analysis: A follow-up procedure to a "significant" MANOVA. *Monograph* presented at the 1994 American Alliance for Health, Physical Education, Recreation and Dance Convention, Denver, CO, April 13.

Witte, R.S. (1985). *Statistics (2nd ed.)*. New York: Holt, Reinhart & Winston.

Winer, B.J., Brown, D.R., & Michels, K.M. (1991). *Statistical principles in experimental design (3rd ed.)* New York: McGraw-Hill.

Index

About the Author

Willian J. Vincent is a professor and chair of the Kinesiology Department at California State University, Northridge (CSUN). He has worked at the school for 36 years and has taught statistics and measurement theory for 30 of those years. In 1995 he received CSUN's Distinguished Teaching Award.

A member of the American Alliance for Health, Physical Education, Recreation and Dance (AAHPERD), Dr. Vincent was elected to serve as the Southwest District representative to the AAHPERD Board of Governors from 1993 to 1995. He is also a member and past president of the Southwest District AAHPERD, a member of the research consortium of AAHPERD, and a member of the southwest chapter of the American College of Sports Medicine. In 1988 he was named the Southwest District AAHPERD Scholar and was given the honor of presenting the keynote address at the 1989 Southwest District AAHPERD convention.

Dr. Vincent is the author of four books and more than 70 professional journal articles. Fifty-one of those articles appeared in referenced journals, including *Research Quarterly for Exercise and Sport*, the *International Journal of Sports Medicine*, and the *Journal of Athletic Training*. He received his doctorate in educational psychology in 1966 from the University of California, Los Angeles. He and his wife Diana have six grown children and live in Northridge, California. In his free time, Dr. Vincent enjoys camping, snow skiing, off-road motorcycle riding, gardening, and reading.

Related Books from Human Kinetics

Research Methods in Physical Activity (Third Edition)

Jerry R. Thomas, EdD, and Jack K. Nelson, EdD
1996 • Hardcover • 504 pp • Item BTHO0481
ISBN 0-88011-481-9 • $49.00 ($73.50 Canadian)

Includes guidelines for conducting original research and essential information for understanding and evaluating existing research. It's a superb text for graduate-level research methods courses and a useful reference book for experienced researchers who want to brush up their skills and become more familiar with new techniques.

Instructor Guide Software for Research Methods in Physical Activity

Windows Item BTHO0608 • *Macintosh* Item BTHO0612 • 1996 • 3-1/2" disk • FREE to course adopters.

The companion instructor guide software provides a course outline, learning activities for each chapter, transparency masters, and a computerized test bank.

Measurement and Evaluation in Human Performance

James R. Morrow, Jr., PhD, Allen W. Jackson, EdD, James G. Disch, PED, and Dale P. Mood, PhD
Book with *Windows* software • 1995 • 416 pp • 3-1/2" disk • Item BMOR0731
ISBN 0-87322-731-X • $49.00 ($73.50 Canadian)

A comprehensive text and software package, this unique textbook includes the powerful MYSTAT, the graphic and data management program that helps students apply important concepts hands-on. The book explains the theories and principles of measurement. Students can use MYSTAT to conduct the tasks assigned in the text as well as create and analyze their own real and hypothetical data sets.

Instructor Guide Software for Measurement and Evaluation in Human Performance

Windows Item BMOR0645 • *Macintosh* Item BMOR0647
1996 • 3 1/2" disk • FREE to course adopters.

To request more information or to order, U.S. customers call 1-800-747-4457, e-mail us at **humank@hkusa.com**, or visit our website at **http://www.humankinetics.com/**. Persons outside the U.S. can contact us via our website or use the appropriate telephone number, postal address, or e-mail address shown in the front of this book.

HUMAN KINETICS
The Information Leader in Physical Activity
P.O. Box 5076, Champaign, IL 61825-5076